Think Bayes

Allen B. Downey

Beijing · Boston · Farnham · Sebastopol · Tokyo

Think Bayes

By Allen B. Downey

Copyright © 2013 Allen B. Downey. All rights reserved.

Printed in the United States of America.

Published by O'Reilly Media, Inc., 1005 Gravenstein Highway North, Sebastopol, CA 95472.

O'Reilly books may be purchased for educational, business, or sales promotional use. Online editions are also available for most titles (*http://safaribooksonline.com*). For more information, contact our corporate/institutional sales department: 800-998-9938 or corporate@oreilly.com.

Editors: Mike Loukides and Ann Spencer
Production Editor: Melanie Yarbrough
Proofreader: Jasmine Kwityn
Indexer: Allen Downey
Interior Designer: David Futato
Cover Designer: Randy Comer
Illustrator: Rebecca Demarest

September, 2013:   First Edition

Revision History for the First Edition
2013-09-10:   First release
2014-02-10:   Second release
2014-08-22:   Third release
2016-06-03:   Fourth release

See *http://oreilly.com/catalog/errata.csp?isbn=9781449370787* for release details.

The O'Reilly logo is a registered trademark of O'Reilly Media, Inc. *Think Bayes*, the cover image of a red striped mullet, and related trade dress are trademarks of O'Reilly Media, Inc.

While the publisher and the authors have used good faith efforts to ensure that the information and instructions contained in this work are accurate, the publisher and the authors disclaim all responsibility for errors or omissions, including without limitation responsibility for damages resulting from the use of or reliance on this work. Use of the information and instructions contained in this work is at your own risk. If any code samples or other technology this work contains or describes is subject to open source licenses or the intellectual property rights of others, it is your responsibility to ensure that your use thereof complies with such licenses and/or rights.

Think Bayes is available under the Creative Commons Attribution-NonCommercial 3.0 Unported License. The author maintains an online version at *http://thinkbayes.com*.

978-1-449-37078-7

[LSI]

Table of Contents

Preface.. ix

1. Bayes's Theorem... 1
    Conditional probability   1
    Conjoint probability   2
    The cookie problem   3
    Bayes's theorem   3
    The diachronic interpretation   5
    The M&M problem   6
    The Monty Hall problem   7
    Discussion   9

2. Computational Statistics.. 11
    Distributions   11
    The cookie problem   12
    The Bayesian framework   13
    The Monty Hall problem   15
    Encapsulating the framework   16
    The M&M problem   16
    Discussion   18
    Exercises   18

3. Estimation.. 19
    The dice problem   19
    The locomotive problem   20
    What about that prior?   23
    An alternative prior   24
    Credible intervals   26

Cumulative distribution functions	26
The German tank problem	27
Discussion	28
Exercises	28

4. More Estimation ... 31
The Euro problem	31
Summarizing the posterior	33
Swamping the priors	33
Optimization	35
The beta distribution	37
Discussion	38
Exercises	39

5. Odds and Addends ... 41
Odds	41
The odds form of Bayes's theorem	42
Oliver's blood	43
Addends	44
Maxima	47
Mixtures	50
Discussion	52

6. Decision Analysis ... 53
The Price is Right problem	53
The prior	54
Probability density functions	55
Representing PDFs	55
Modeling the contestants	58
Likelihood	60
Update	61
Optimal bidding	62
Discussion	65

7. Prediction ... 67
The Boston Bruins problem	67
Poisson processes	68
The posteriors	69
The distribution of goals	70
The probability of winning	72
Sudden death	73
Discussion	75

| Exercises | 76 |

8. Observer Bias
The Red Line problem	79
The model	79
Wait times	81
Predicting wait times	84
Estimating the arrival rate	87
Incorporating uncertainty	89
Decision analysis	91
Discussion	93
Exercises	94

9. Two Dimensions
Paintball	95
The suite	96
Trigonometry	97
Likelihood	99
Joint distributions	100
Conditional distributions	101
Credible intervals	102
Discussion	104
Exercises	105

10. Approximate Bayesian Computation
The Variability Hypothesis	107
Mean and standard deviation	108
Update	110
The posterior distribution of CV	110
Underflow	111
Log-likelihood	113
A little optimization	113
ABC	115
Robust estimation	116
Who is more variable?	118
Discussion	121
Exercises	121

11. Hypothesis Testing
Back to the Euro problem	123
Making a fair comparison	124
The triangle prior	126

Discussion	126
Exercises	127

12. Evidence 129
Interpreting SAT scores	129
The scale	130
The prior	130
Posterior	132
A better model	134
Calibration	136
Posterior distribution of efficacy	137
Predictive distribution	139
Discussion	140

13. Simulation 143
The Kidney Tumor problem	143
A simple model	145
A more general model	146
Implementation	148
Caching the joint distribution	149
Conditional distributions	150
Serial Correlation	152
Discussion	155

14. A Hierarchical Model 157
The Geiger counter problem	157
Start simple	158
Make it hierarchical	159
A little optimization	160
Extracting the posteriors	161
Discussion	162
Exercises	163

15. Dealing with Dimensions 165
Belly button bacteria	165
Lions and tigers and bears	166
The hierarchical version	168
Random sampling	170
Optimization	172
Collapsing the hierarchy	172
One more problem	175
We're not done yet	176

The belly button data	178
Predictive distributions	181
Joint posterior	185
Coverage	186
Discussion	188

Index **191**

Preface

My theory, which is mine

The premise of this book, and the other books in the *Think X* series, is that if you know how to program, you can use that skill to learn other topics.

Most books on Bayesian statistics use mathematical notation and present ideas in terms of mathematical concepts like calculus. This book uses Python code instead of math, and discrete approximations instead of continuous mathematics. As a result, what would be an integral in a math book becomes a summation, and most operations on probability distributions are simple loops.

I think this presentation is easier to understand, at least for people with programming skills. It is also more general, because when we make modeling decisions, we can choose the most appropriate model without worrying too much about whether the model lends itself to conventional analysis.

Also, it provides a smooth development path from simple examples to real-world problems. Chapter 3 is a good example. It starts with a simple example involving dice, one of the staples of basic probability. From there it proceeds in small steps to the locomotive problem, which I borrowed from Mosteller's *Fifty Challenging Problems in Probability with Solutions*, and from there to the German tank problem, a famously successful application of Bayesian methods during World War II.

Modeling and approximation

Most chapters in this book are motivated by a real-world problem, so they involve some degree of modeling. Before we can apply Bayesian methods (or any other analysis), we have to make decisions about which parts of the real-world system to include in the model and which details we can abstract away.

For example, in Chapter 7, the motivating problem is to predict the winner of a hockey game. I model goal-scoring as a Poisson process, which implies that a goal is equally likely at any point in the game. That is not exactly true, but it is probably a good enough model for most purposes.

In Chapter 12 the motivating problem is interpreting SAT scores (the SAT is a standardized test used for college admissions in the United States). I start with a simple model that assumes that all SAT questions are equally difficult, but in fact the designers of the SAT deliberately include some questions that are relatively easy and some that are relatively hard. I present a second model that accounts for this aspect of the design, and show that it doesn't have a big effect on the results after all.

I think it is important to include modeling as an explicit part of problem solving because it reminds us to think about modeling errors (that is, errors due to simplifications and assumptions of the model).

Many of the methods in this book are based on discrete distributions, which makes some people worry about numerical errors. But for real-world problems, numerical errors are almost always smaller than modeling errors.

Furthermore, the discrete approach often allows better modeling decisions, and I would rather have an approximate solution to a good model than an exact solution to a bad model.

On the other hand, continuous methods sometimes yield performance advantages—for example by replacing a linear- or quadratic-time computation with a constant-time solution.

So I recommend a general process with these steps:

1. While you are exploring a problem, start with simple models and implement them in code that is clear, readable, and demonstrably correct. Focus your attention on good modeling decisions, not optimization.

2. Once you have a simple model working, identify the biggest sources of error. You might need to increase the number of values in a discrete approximation, or increase the number of iterations in a Monte Carlo simulation, or add details to the model.

3. If the performance of your solution is good enough for your application, you might not have to do any optimization. But if you do, there are two approaches to consider. You can review your code and look for optimizations; for example, if you cache previously computed results you might be able to avoid redundant computation. Or you can look for analytic methods that yield computational shortcuts.

One benefit of this process is that Steps 1 and 2 tend to be fast, so you can explore several alternative models before investing heavily in any of them.

Another benefit is that if you get to Step 3, you will be starting with a reference implementation that is likely to be correct, which you can use for regression testing (that is, checking that the optimized code yields the same results, at least approximately).

Working with the code

The code and sound samples used in this book are available from *https://github.com/AllenDowney/ThinkBayes*. Git is a version control system that allows you to keep track of the files that make up a project. A collection of files under Git's control is called a "repository". GitHub is a hosting service that provides storage for Git repositories and a convenient web interface.

The GitHub homepage for my repository provides several ways to work with the code:

- You can create a copy of my repository on GitHub by pressing the Fork button. If you don't already have a GitHub account, you'll need to create one. After forking, you'll have your own repository on GitHub that you can use to keep track of code you write while working on this book. Then you can clone the repo, which means that you copy the files to your computer.
- Or you could clone my repository. You don't need a GitHub account to do this, but you won't be able to write your changes back to GitHub.
- If you don't want to use Git at all, you can download the files in a Zip file using the button in the lower-right corner of the GitHub page.

The code for the first edition of the book works with Python 2. If you are using Python 3, you might want to use the updated code in *https://github.com/AllenDowney/ThinkBayes2* instead.

I developed this book using Anaconda from Continuum Analytics, which is a free Python distribution that includes all the packages you'll need to run the code (and lots more). I found Anaconda easy to install. By default it does a user-level installation, not system-level, so you don't need administrative privileges. You can download Anaconda from *http://continuum.io/downloads*.

If you don't want to use Anaconda, you will need the following packages:

- NumPy for basic numerical computation, *http://www.numpy.org/*;
- SciPy for scientific computation, *http://www.scipy.org/*;
- matplotlib for visualization, *http://matplotlib.org/*.

Although these are commonly used packages, they are not included with all Python installations, and they can be hard to install in some environments. If you have trouble installing them, I recommend using Anaconda or one of the other Python distributions that include these packages.

Many of the examples in this book use classes and functions defined in `thinkbayes.py`. Some of them also use `thinkplot.py`, which provides wrappers for some of the functions in `pyplot`, which is part of `matplotlib`.

Code style

Experienced Python programmers will notice that the code in this book does not comply with PEP 8, which is the most common style guide for Python (*http://www.python.org/dev/peps/pep-0008/*).

Specifically, PEP 8 calls for lowercase function names with underscores between words, `like_this`. In this book and the accompanying code, function and method names begin with a capital letter and use camel case, `LikeThis`.

I broke this rule because I developed some of the code while I was a Visiting Scientist at Google, so I followed the Google style guide, which deviates from PEP 8 in a few places. Once I got used to Google style, I found that I liked it. And at this point, it would be too much trouble to change.

Also on the topic of style, I write "Bayes's theorem" with an *s* after the apostrophe, which is preferred in some style guides and deprecated in others. I don't have a strong preference. I had to choose one, and this is the one I chose.

And finally one typographical note: throughout the book, I use PMF and CDF for the mathematical concept of a probability mass function or cumulative distribution function, and Pmf and Cdf to refer to the Python objects I use to represent them.

Prerequisites

There are several excellent modules for doing Bayesian statistics in Python, including `pymc` and OpenBUGS. I chose not to use them for this book because you need a fair amount of background knowledge to get started with these modules, and I want to keep the prerequisites minimal. If you know Python and a little bit about probability, you are ready to start this book.

Chapter 1 is about probability and Bayes's theorem; it has no code. Chapter 2 introduces `Pmf`, a thinly disguised Python dictionary I use to represent a probability mass function (PMF). Then Chapter 3 introduces `Suite`, a kind of Pmf that provides a framework for doing Bayesian updates. And that's just about all there is to it.

Well, almost. In some of the later chapters, I use analytic distributions including the Gaussian (normal) distribution, the exponential and Poisson distributions, and the beta distribution. In Chapter 15 I break out the less-common Dirichlet distribution, but I explain it as I go along. If you are not familiar with these distributions, you can read about them on Wikipedia. You could also read the companion to this book, *Think Stats*, or an introductory statistics book (although I'm afraid most of them take a mathematical approach that is not particularly helpful for practical purposes).

Conventions Used in This Book

The following typographical conventions are used in this book:

Italic
: Indicates new terms, URLs, email addresses, filenames, and file extensions.

`Constant width`
: Used for program listings, as well as within paragraphs to refer to program elements such as variable or function names, databases, data types, environment variables, statements, and keywords.

`Constant width bold`
: Shows commands or other text that should be typed literally by the user.

`Constant width italic`
: Shows text that should be replaced with user-supplied values or by values determined by context.

This icon signifies a tip, suggestion, or general note.

This icon indicates a warning or caution.

Safari® Books Online

Safari Books Online (*www.safaribooksonline.com*) is an on-demand digital library that delivers expert content in both book and video form from the world's leading authors in technology and business.

Technology professionals, software developers, web designers, and business and creative professionals use Safari Books Online as their primary resource for research, problem solving, learning, and certification training.

Safari Books Online offers a range of plans and pricing for enterprise, government, education, and individuals. Members have access to thousands of books, training videos, and prepublication manuscripts in one fully searchable database from publishers like O'Reilly Media, Prentice Hall Professional, Addison-Wesley Professional, Microsoft Press, Sams, Que, Peachpit Press, Focal Press, Cisco Press, John Wiley & Sons, Syngress, Morgan Kaufmann, IBM Redbooks, Packt, Adobe Press, FT Press, Apress, Manning, New Riders, McGraw-Hill, Jones & Bartlett, Course Technology, and hundreds more. For more information about Safari Books Online, please visit us online.

How to Contact Us

Please address comments and questions concerning this book to the publisher:

> O'Reilly Media, Inc.
> 1005 Gravenstein Highway North
> Sebastopol, CA 95472
> 800-998-9938 (in the United States or Canada)
> 707-829-0515 (international or local)
> 707-829-0104 (fax)

We have a web page for this book, where we list errata, examples, and any additional information. You can access this page at *http://oreil.ly/think-bayes*.

To comment or ask technical questions about this book, send email to *bookquestions@oreilly.com*.

For more information about our books, courses, conferences, and news, see our website at *http://www.oreilly.com*.

Find us on Facebook: *http://facebook.com/oreilly*

Follow us on Twitter: *http://twitter.com/oreillymedia*

Watch us on YouTube: *http://www.youtube.com/oreillymedia*

Contributor List

If you have a suggestion or correction, please send email to *downey@allendowney.com*. If I make a change based on your feedback, I will add you to the contributor list (unless you ask to be omitted).

If you include at least part of the sentence the error appears in, that makes it easy for me to search. Page and section numbers are fine, too, but not as easy to work with. Thanks!

- First, I have to acknowledge David MacKay's excellent book, *Information Theory, Inference, and Learning Algorithms*, which is where I first came to understand Bayesian methods. With his permission, I use several problems from his book as examples.
- This book also benefited from my interactions with Sanjoy Mahajan, especially in fall 2012, when I audited his class on Bayesian Inference at Olin College.
- I wrote parts of this book during project nights with the Boston Python User Group, so I would like to thank them for their company and pizza.
- Jonathan Edwards sent in the first typo.
- George Purkins found a markup error.
- Olivier Yiptong sent several helpful suggestions.
- Yuriy Pasichnyk found several errors.
- Kristopher Overholt sent a long list of corrections and suggestions.
- Robert Marcus found a misplaced *i*.
- Max Hailperin suggested a clarification in Chapter 1.
- Markus Dobler pointed out that drawing cookies from a bowl with replacement is an unrealistic scenario.
- Tom Pollard and Paul A. Giannaros spotted a version problem with some of the numbers in the train example.
- Ram Limbu found a typo and suggested a clarification.
- In spring 2013, students in my class, Computational Bayesian Statistics, made many helpful corrections and suggestions: Kai Austin, Claire Barnes, Kari Bender, Rachel Boy, Kat Mendoza, Arjun Iyer, Ben Kroop, Nathan Lintz, Kyle McConnaughay, Alec Radford, Brendan Ritter, and Evan Simpson.
- Greg Marra and Matt Aasted helped me clarify the discussion of *The Price is Right* problem.
- Marcus Ogren pointed out that the original statement of the locomotive problem was ambiguous.
- Jasmine Kwityn and Dan Fauxsmith at O'Reilly Media proofread the book and found many opportunities for improvement.

CHAPTER 1
Bayes's Theorem

Conditional probability

The fundamental idea behind all Bayesian statistics is Bayes's theorem, which is surprisingly easy to derive, provided that you understand conditional probability. So we'll start with probability, then conditional probability, then Bayes's theorem, and on to Bayesian statistics.

A probability is a number between 0 and 1 (including both) that represents a degree of belief in a fact or prediction. The value 1 represents certainty that a fact is true, or that a prediction will come true. The value 0 represents certainty that the fact is false.

Intermediate values represent degrees of certainty. The value 0.5, often written as 50%, means that a predicted outcome is as likely to happen as not. For example, the probability that a tossed coin lands face up is very close to 50%.

A conditional probability is a probability based on some background information. For example, I want to know the probability that I will have a heart attack in the next year. According to the CDC, "Every year about 785,000 Americans have a first coronary attack (*http://www.cdc.gov/heartdisease/facts.htm*)."

The U.S. population is about 311 million, so the probability that a randomly chosen American will have a heart attack in the next year is roughly 0.3%.

But I am not a randomly chosen American. Epidemiologists have identified many factors that affect the risk of heart attacks; depending on those factors, my risk might be higher or lower than average.

I am male, 45 years old, and I have borderline high cholesterol. Those factors increase my chances. However, I have low blood pressure and I don't smoke, and those factors decrease my chances.

Plugging everything into the online calculator at *http://cvdrisk.nhlbi.nih.gov/calculator.asp*, I find that my risk of a heart attack in the next year is about 0.2%, less than the national average. That value is a conditional probability, because it is based on a number of factors that make up my "condition."

The usual notation for conditional probability is p(A|B), which is the probability of A given that B is true. In this example, A represents the prediction that I will have a heart attack in the next year, and B is the set of conditions I listed.

Conjoint probability

Conjoint probability is a fancy way to say the probability that two things are true. I write p(A and B) to mean the probability that A and B are both true.

If you learned about probability in the context of coin tosses and dice, you might have learned the following formula:

p(A and B) = p(A) p(B)   WARNING: not always true

For example, if I toss two coins, and A means the first coin lands face up, and B means the second coin lands face up, then p(A) = p(B) = 0.5, and sure enough, p(A and B) = p(A) p(B) = 0.25.

But this formula only works because in this case A and B are independent; that is, knowing the outcome of the first event does not change the probability of the second. Or, more formally, p(B|A) = p(B).

Here is a different example where the events are not independent. Suppose that A means that it rains today and B means that it rains tomorrow. If I know that it rained today, it is more likely that it will rain tomorrow, so p(B|A) > p(B).

In general, the probability of a conjunction is

p(A and B) = p(A) p(B|A)

for any A and B. So if the chance of rain on any given day is 0.5, the chance of rain on two consecutive days is not 0.25, but probably a bit higher.

The cookie problem

We'll get to Bayes's theorem soon, but I want to motivate it with an example called the cookie problem.[1] Suppose there are two bowls of cookies. Bowl 1 contains 30 vanilla cookies and 10 chocolate cookies. Bowl 2 contains 20 of each.

Now suppose you choose one of the bowls at random and, without looking, select a cookie at random. The cookie is vanilla. What is the probability that it came from Bowl 1?

This is a conditional probability; we want p(Bowl 1|vanilla), but it is not obvious how to compute it. If I asked a different question—the probability of a vanilla cookie given Bowl 1—it would be easy:

   p(vanilla|Bowl 1) = 3/4

Sadly, p(A|B) is *not* the same as p(B|A), but there is a way to get from one to the other: Bayes's theorem.

Bayes's theorem

At this point we have everything we need to derive Bayes's theorem. We'll start with the observation that conjunction is commutative; that is

   p(A and B) = p(B and A)

for any events A and B.

Next, we write the probability of a conjunction:

   p(A and B) = p(A) p(B|A)

Since we have not said anything about what A and B mean, they are interchangeable. Interchanging them yields

   p(B and A) = p(B) p(A|B)

That's all we need. Pulling those pieces together, we get

   p(B) p(A|B) = p(A) p(B|A)

---

[1] Based on an example from *http://en.wikipedia.org/wiki/Bayes'_theorem* that is no longer there.

Which means there are two ways to compute the conjunction. If you have p(A), you multiply by the conditional probability p(B|A). Or you can do it the other way around; if you know p(B), you multiply by p(A|B). Either way you should get the same thing.

Finally we can divide through by p(B):

$$p(A|B) = \frac{p(A)\ p(B|A)}{p(B)}$$

And that's Bayes's theorem! It might not look like much, but it turns out to be surprisingly powerful.

For example, we can use it to solve the cookie problem. I'll write B_1 for the hypothesis that the cookie came from Bowl 1 and V for the vanilla cookie. Plugging in Bayes's theorem we get

$$p(B_1|V) = \frac{p(B_1)\ p(V|B_1)}{p(V)}$$

The term on the left is what we want: the probability of Bowl 1, given that we chose a vanilla cookie. The terms on the right are:

- $p(B_1)$: This is the probability that we chose Bowl 1, unconditioned by what kind of cookie we got. Since the problem says we chose a bowl at random, we can assume $p(B_1) = 1/2$.
- $p(V|B_1)$: This is the probability of getting a vanilla cookie from Bowl 1, which is 3/4.
- $p(V)$: This is the probability of drawing a vanilla cookie from either bowl. Since we had an equal chance of choosing either bowl and the bowls contain the same number of cookies, we had the same chance of choosing any cookie. Between the two bowls there are 50 vanilla and 30 chocolate cookies, so $p(V) = 5/8$.

Putting it together, we have

$$p(B_1|V) = \frac{(1/2)\ (3/4)}{5/8}$$

which reduces to 3/5. So the vanilla cookie is evidence in favor of the hypothesis that we chose Bowl 1, because vanilla cookies are more likely to come from Bowl 1.

This example demonstrates one use of Bayes's theorem: it provides a strategy to get from p(B|A) to p(A|B). This strategy is useful in cases, like the cookie problem, where it is easier to compute the terms on the right side of Bayes's theorem than the term on the left.

The diachronic interpretation

There is another way to think of Bayes's theorem: it gives us a way to update the probability of a hypothesis, *H*, in light of some body of data, *D*.

This way of thinking about Bayes's theorem is called the **diachronic interpretation**. "Diachronic" means that something is happening over time; in this case the probability of the hypotheses changes, over time, as we see new data.

Rewriting Bayes's theorem with *H* and *D* yields:

$$p(H|D) = \frac{p(H)\,p(D|H)}{p(D)}$$

In this interpretation, each term has a name:

- $p(H)$ is the probability of the hypothesis before we see the data, called the prior probability, or just **prior**.
- $p(H|D)$ is what we want to compute, the probability of the hypothesis after we see the data, called the **posterior**.
- $p(D|H)$ is the probability of the data under the hypothesis, called the **likelihood**.
- $p(D)$ is the probability of the data under any hypothesis, called the **normalizing constant**.

Sometimes we can compute the prior based on background information. For example, the cookie problem specifies that we choose a bowl at random with equal probability.

In other cases the prior is subjective; that is, reasonable people might disagree, either because they use different background information or because they interpret the same information differently.

The likelihood is usually the easiest part to compute. In the cookie problem, if we know which bowl the cookie came from, we find the probability of a vanilla cookie by counting.

The normalizing constant can be tricky. It is supposed to be the probability of seeing the data under any hypothesis at all, but in the most general case it is hard to nail down what that means.

Most often we simplify things by specifying a set of hypotheses that are

Mutually exclusive:
   At most one hypothesis in the set can be true, and

Collectively exhaustive:
   There are no other possibilities; at least one of the hypotheses has to be true.

I use the word **suite** for a set of hypotheses that has these properties.

In the cookie problem, there are only two hypotheses—the cookie came from Bowl 1 or Bowl 2—and they are mutually exclusive and collectively exhaustive.

In that case we can compute p(D) using the law of total probability, which says that if there are two exclusive ways that something might happen, you can add up the probabilities like this:

$$p(D) = p(B_1)\, p(D|B_1) + p(B_2)\, p(D|B_2)$$

Plugging in the values from the cookie problem, we have

$$p(D) = (1/2)\,(3/4) + (1/2)\,(1/2) = 5/8$$

which is what we computed earlier by mentally combining the two bowls.

The M&M problem

M&M's are small candy-coated chocolates that come in a variety of colors. Mars, Inc., which makes M&M's, changes the mixture of colors from time to time.

In 1995, they introduced blue M&M's. Before then, the color mix in a bag of plain M&M's was 30% Brown, 20% Yellow, 20% Red, 10% Green, 10% Orange, 10% Tan. Afterward it was 24% Blue, 20% Green, 16% Orange, 14% Yellow, 13% Red, 13% Brown.

Suppose a friend of mine has two bags of M&M's, and he tells me that one is from 1994 and one from 1996. He won't tell me which is which, but he gives me one M&M from each bag. One is yellow and one is green. What is the probability that the yellow one came from the 1994 bag?

This problem is similar to the cookie problem, with the twist that I draw one sample from each bowl/bag. This problem also gives me a chance to demonstrate the table method, which is useful for solving problems like this on paper. In the next chapter we will solve them computationally.

The first step is to enumerate the hypotheses. The bag the yellow M&M came from I'll call Bag 1; I'll call the other Bag 2. So the hypotheses are:

- A: Bag 1 is from 1994, which implies that Bag 2 is from 1996.
- B: Bag 1 is from 1996 and Bag 2 from 1994.

Now we construct a table with a row for each hypothesis and a column for each term in Bayes's theorem:

	Prior $p(H)$	Likelihood $p(D\|H)$	$p(H)\,p(D\|H)$	Posterior $p(H\|D)$
A	1/2	(20)(20)	200	20/27
B	1/2	(10)(14)	70	7/27

The first column has the priors. Based on the statement of the problem, it is reasonable to choose $p(A) = p(B) = 1/2$.

The second column has the likelihoods, which follow from the information in the problem. For example, if A is true, the yellow M&M came from the 1994 bag with probability 20%, and the green came from the 1996 bag with probability 20%. Because the selections are independent, we get the conjoint probability by multiplying.

The third column is just the product of the previous two. The sum of this column, 270, is the normalizing constant. To get the last column, which contains the posteriors, we divide the third column by the normalizing constant.

That's it. Simple, right?

Well, you might be bothered by one detail. I write $p(D|H)$ in terms of percentages, not probabilities, which means it is off by a factor of 10,000. But that cancels out when we divide through by the normalizing constant, so it doesn't affect the result.

When the set of hypotheses is mutually exclusive and collectively exhaustive, you can multiply the likelihoods by any factor, if it is convenient, as long as you apply the same factor to the entire column.

The Monty Hall problem

The Monty Hall problem might be the most contentious question in the history of probability. The scenario is simple, but the correct answer is so counterintuitive that many people just can't accept it, and many smart people have embarrassed themselves not just by getting it wrong but by arguing the wrong side, aggressively, in public.

Monty Hall was the original host of the game show *Let's Make a Deal*. The Monty Hall problem is based on one of the regular games on the show. If you are on the show, here's what happens:

- Monty shows you three closed doors and tells you that there is a prize behind each door: one prize is a car, the other two are less valuable prizes like peanut butter and fake finger nails. The prizes are arranged at random.
- The object of the game is to guess which door has the car. If you guess right, you get to keep the car.
- You pick a door, which we will call Door A. We'll call the other doors B and C.
- Before opening the door you chose, Monty increases the suspense by opening either Door B or C, whichever does not have the car. (If the car is actually behind Door A, Monty can safely open B or C, so he chooses one at random.)
- Then Monty offers you the option to stick with your original choice or switch to the one remaining unopened door.

The question is, should you "stick" or "switch" or does it make no difference?

Most people have the strong intuition that it makes no difference. There are two doors left, they reason, so the chance that the car is behind Door A is 50%.

But that is wrong. In fact, the chance of winning if you stick with Door A is only 1/3; if you switch, your chances are 2/3.

By applying Bayes's theorem, we can break this problem into simple pieces, and maybe convince ourselves that the correct answer is, in fact, correct.

To start, we should make a careful statement of the data. In this case D consists of two parts: Monty chooses Door B *and* there is no car there.

Next we define three hypotheses: A, B, and C represent the hypothesis that the car is behind Door A, Door B, or Door C. Again, let's apply the table method:

	Prior $p(H)$	Likelihood $p(D\|H)$	$p(H)\, p(D\|H)$	Posterior $p(H\|D)$
A	1/3	1/2	1/6	1/3
B	1/3	0	0	0
C	1/3	1	1/3	2/3

Filling in the priors is easy because we are told that the prizes are arranged at random, which suggests that the car is equally likely to be behind any door.

Figuring out the likelihoods takes some thought, but with reasonable care we can be confident that we have it right:

- If the car is actually behind A, Monty could safely open Doors B or C. So the probability that he chooses B is 1/2. And since the car is actually behind A, the probability that the car is not behind B is 1.
- If the car is actually behind B, Monty has to open door C, so the probability that he opens door B is 0.
- Finally, if the car is behind Door C, Monty opens B with probability 1 and finds no car there with probability 1.

Now the hard part is over; the rest is just arithmetic. The sum of the third column is 1/2. Dividing through yields $p(A|D) = 1/3$ and $p(C|D) = 2/3$. So you are better off switching.

There are many variations of the Monty Hall problem. One of the strengths of the Bayesian approach is that it generalizes to handle these variations.

For example, suppose that Monty always chooses B if he can, and only chooses C if he has to (because the car is behind B). In that case the revised table is:

	Prior $p(H)$	Likelihood $p(D|H)$	$p(H)\,p(D|H)$	Posterior $p(H|D)$
A	1/3	1	1/3	1/2
B	1/3	0	0	0
C	1/3	1	1/3	1/2

The only change is $p(D|A)$. If the car is behind A, Monty can choose to open B or C. But in this variation he always chooses B, so $p(D|A) = 1$.

As a result, the likelihoods are the same for A and C, and the posteriors are the same: $p(A|D) = p(C|D) = 1/2$. In this case, the fact that Monty chose B reveals no information about the location of the car, so it doesn't matter whether the contestant sticks or switches.

On the other hand, if he had opened C, we would know $p(B|D) = 1$.

I included the Monty Hall problem in this chapter because I think it is fun, and because Bayes's theorem makes the complexity of the problem a little more manageable. But it is not a typical use of Bayes's theorem, so if you found it confusing, don't worry!

Discussion

For many problems involving conditional probability, Bayes's theorem provides a divide-and-conquer strategy. If $p(A|B)$ is hard to compute, or hard to measure experimentally, check whether it might be easier to compute the other terms in Bayes's theorem, $p(B|A)$, $p(A)$ and $p(B)$.

If the Monty Hall problem is your idea of fun, I have collected a number of similar problems in an article called "All your Bayes are belong to us," which you can read at *http://allendowney.blogspot.com/2011/10/all-your-bayes-are-belong-to-us.html*.

CHAPTER 2
Computational Statistics

Distributions

In statistics a **distribution** is a set of values and their corresponding probabilities.

For example, if you roll a six-sided die, the set of possible values is the numbers 1 to 6, and the probability associated with each value is 1/6.

As another example, you might be interested in how many times each word appears in common English usage. You could build a distribution that includes each word and how many times it appears.

To represent a distribution in Python, you could use a dictionary that maps from each value to its probability. I have written a class called Pmf that uses a Python dictionary in exactly that way, and provides a number of useful methods. I called the class Pmf in reference to a **probability mass function**, which is a way to represent a distribution mathematically.

Pmf is defined in a Python module I wrote to accompany this book, thinkbayes.py. You can download it from *http://thinkbayes.com/thinkbayes.py*. For more information see "Working with the code" on page xi.

To use Pmf you can import it like this:

```
from thinkbayes import Pmf
```

The following code builds a Pmf to represent the distribution of outcomes for a six-sided die:

```
pmf = Pmf()
for x in [1,2,3,4,5,6]:
    pmf.Set(x, 1/6.0)
```

Pmf creates an empty Pmf with no values. The Set method sets the probability associated with each value to 1/6.

Here's another example that counts the number of times each word appears in a sequence:

```
pmf = Pmf()
for word in word_list:
    pmf.Incr(word, 1)
```

Incr increases the "probability" associated with each word by 1. If a word is not already in the Pmf, it is added.

I put "probability" in quotes because in this example, the probabilities are not normalized; that is, they do not add up to 1. So they are not true probabilities.

But in this example the word counts are proportional to the probabilities. So after we count all the words, we can compute probabilities by dividing through by the total number of words. Pmf provides a method, Normalize, that does exactly that:

```
pmf.Normalize()
```

Once you have a Pmf object, you can ask for the probability associated with any value:

```
print pmf.Prob('the')
```

And that would print the frequency of the word "the" as a fraction of the words in the list.

Pmf uses a Python dictionary to store the values and their probabilities, so the values in the Pmf can be any hashable type. The probabilities can be any numerical type, but they are usually floating-point numbers (type float).

The cookie problem

In the context of Bayes's theorem, it is natural to use a Pmf to map from each hypothesis to its probability. In the cookie problem, the hypotheses are B_1 and B_2. In Python, I represent them with strings:

```
pmf = Pmf()
pmf.Set('Bowl 1', 0.5)
pmf.Set('Bowl 2', 0.5)
```

This distribution, which contains the priors for each hypothesis, is called (wait for it) the **prior distribution**.

To update the distribution based on new data (the vanilla cookie), we multiply each prior by the corresponding likelihood. The likelihood of drawing a vanilla cookie from Bowl 1 is 3/4. The likelihood for Bowl 2 is 1/2.

```
pmf.Mult('Bowl 1', 0.75)
pmf.Mult('Bowl 2', 0.5)
```

Mult does what you would expect. It gets the probability for the given hypothesis and multiplies by the given likelihood.

After this update, the distribution is no longer normalized, but because these hypotheses are mutually exclusive and collectively exhaustive, we can **renormalize**:

```
pmf.Normalize()
```

The result is a distribution that contains the posterior probability for each hypothesis, which is called (wait now) the **posterior distribution**.

Finally, we can get the posterior probability for Bowl 1:

```
print pmf.Prob('Bowl 1')
```

And the answer is 0.6. You can download this example from *http://thinkbayes.com/cookie.py*. For more information see "Working with the code" on page xi.

The Bayesian framework

Before we go on to other problems, I want to rewrite the code from the previous section to make it more general. First I'll define a class to encapsulate the code related to this problem:

```
class Cookie(Pmf):

    def __init__(self, hypos):
        Pmf.__init__(self)
        for hypo in hypos:
            self.Set(hypo, 1)
        self.Normalize()
```

A Cookie object is a Pmf that maps from hypotheses to their probabilities. The __init__ method gives each hypothesis the same prior probability. As in the previous section, there are two hypotheses:

```
hypos = ['Bowl 1', 'Bowl 2']
pmf = Cookie(hypos)
```

Cookie provides an Update method that takes data as a parameter and updates the probabilities:

```
def Update(self, data):
    for hypo in self.Values():
        like = self.Likelihood(data, hypo)
        self.Mult(hypo, like)
    self.Normalize()
```

`Update` loops through each hypothesis in the suite and multiplies its probability by the likelihood of the data under the hypothesis, which is computed by `Likelihood`:

```
mixes = {
    'Bowl 1':dict(vanilla=0.75, chocolate=0.25),
    'Bowl 2':dict(vanilla=0.5, chocolate=0.5),
    }

def Likelihood(self, data, hypo):
    mix = self.mixes[hypo]
    like = mix[data]
    return like
```

`Likelihood` uses `mixes`, which is a dictionary that maps from the name of a bowl to the mix of cookies in the bowl.

Here's what the update looks like:

```
pmf.Update('vanilla')
```

And then we can print the posterior probability of each hypothesis:

```
for hypo, prob in pmf.Items():
    print hypo, prob
```

The result is

```
Bowl 1 0.6
Bowl 2 0.4
```

which is the same as what we got before. This code is more complicated than what we saw in the previous section. One advantage is that it generalizes to the case where we draw more than one cookie from the same bowl (with replacement):

```
dataset = ['vanilla', 'chocolate', 'vanilla']
for data in dataset:
    pmf.Update(data)
```

The other advantage is that it provides a framework for solving many similar problems. In the next section we'll solve the Monty Hall problem computationally and then see what parts of the framework are the same.

The code in this section is available from *http://thinkbayes.com/cookie2.py*. For more information see "Working with the code" on page xi.

The Monty Hall problem

To solve the Monty Hall problem, I'll define a new class:

```
class Monty(Pmf):

    def __init__(self, hypos):
        Pmf.__init__(self)
        for hypo in hypos:
            self.Set(hypo, 1)
        self.Normalize()
```

So far `Monty` and `Cookie` are exactly the same. And the code that creates the Pmf is the same, too, except for the names of the hypotheses:

```
hypos = 'ABC'
pmf = Monty(hypos)
```

Calling `Update` is pretty much the same:

```
data = 'B'
pmf.Update(data)
```

And the implementation of `Update` is exactly the same:

```
def Update(self, data):
    for hypo in self.Values():
        like = self.Likelihood(data, hypo)
        self.Mult(hypo, like)
    self.Normalize()
```

The only part that requires some work is `Likelihood`:

```
def Likelihood(self, data, hypo):
    if hypo == data:
        return 0
    elif hypo == 'A':
        return 0.5
    else:
        return 1
```

Finally, printing the results is the same:

```
for hypo, prob in pmf.Items():
    print hypo, prob
```

And the answer is

```
A 0.333333333333
B 0.0
C 0.666666666667
```

In this example, writing `Likelihood` is a little complicated, but the framework of the Bayesian update is simple. The code in this section is available from http://thinkbayes.com/monty.py. For more information see "Working with the code" on page xi.

Encapsulating the framework

Now that we see what elements of the framework are the same, we can encapsulate them in an object—a `Suite` is a `Pmf` that provides `__init__`, `Update`, and `Print`:

```python
class Suite(Pmf):
    """Represents a suite of hypotheses and their probabilities."""

    def __init__(self, hypo=tuple()):
        """Initializes the distribution."""

    def Update(self, data):
        """Updates each hypothesis based on the data."""

    def Print(self):
        """Prints the hypotheses and their probabilities."""
```

The implementation of `Suite` is in `thinkbayes.py`. To use `Suite`, you should write a class that inherits from it and provides `Likelihood`. For example, here is the solution to the Monty Hall problem rewritten to use `Suite`:

```python
from thinkbayes import Suite

class Monty(Suite):

    def Likelihood(self, data, hypo):
        if hypo == data:
            return 0
        elif hypo == 'A':
            return 0.5
        else:
            return 1
```

And here's the code that uses this class:

```python
suite = Monty('ABC')
suite.Update('B')
suite.Print()
```

You can download this example from *http://thinkbayes.com/monty2.py*. For more information see "Working with the code" on page xi.

The M&M problem

We can use the `Suite` framework to solve the M&M problem. Writing the `Likelihood` function is tricky, but everything else is straightforward.

First I need to encode the color mixes from before and after 1995:

```
mix94 = dict(brown=30,
             yellow=20,
             red=20,
             green=10,
             orange=10,
             tan=10)

mix96 = dict(blue=24,
             green=20,
             orange=16,
             yellow=14,
             red=13,
             brown=13)
```

Then I have to encode the hypotheses:

```
hypoA = dict(bag1=mix94, bag2=mix96)
hypoB = dict(bag1=mix96, bag2=mix94)
```

hypoA represents the hypothesis that Bag 1 is from 1994 and Bag 2 from 1996. hypoB is the other way around.

Next I map from the name of the hypothesis to the representation:

```
hypotheses = dict(A=hypoA, B=hypoB)
```

And finally I can write Likelihood. In this case the hypothesis, hypo, is a string, either A or B. The data is a tuple that specifies a bag and a color.

```
def Likelihood(self, data, hypo):
    bag, color = data
    mix = self.hypotheses[hypo][bag]
    like = mix[color]
    return like
```

Here's the code that creates the suite and updates it:

```
suite = M_and_M('AB')

suite.Update(('bag1', 'yellow'))
suite.Update(('bag2', 'green'))

suite.Print()
```

And here's the result:

```
A 0.740740740741
B 0.259259259259
```

The posterior probability of A is approximately 20/27, which is what we got before.

The code in this section is available from *http://thinkbayes.com/m_and_m.py*. For more information see "Working with the code" on page xi.

Discussion

This chapter presents the Suite class, which encapsulates the Bayesian update framework.

Suite is an **abstract type**, which means that it defines the interface a Suite is supposed to have, but does not provide a complete implementation. The Suite interface includes Update and Likelihood, but the Suite class only provides an implementation of Update, not Likelihood.

A **concrete type** is a class that extends an abstract parent class and provides an implementation of the missing methods. For example, Monty extends Suite, so it inherits Update and provides Likelihood.

If you are familiar with design patterns, you might recognize this as an example of the template method pattern. You can read about this pattern at *http://en.wikipedia.org/wiki/Template_method_pattern*.

Most of the examples in the following chapters follow the same pattern; for each problem we define a new class that extends Suite, inherits Update, and provides Likelihood. In a few cases we override Update, usually to improve performance.

Exercises

Exercise 2-1.

In "The Bayesian framework" on page 13 I said that the solution to the cookie problem generalizes to the case where we draw multiple cookies with replacement.

But in the more likely scenario where we eat the cookies we draw, the likelihood of each draw depends on the previous draws.

Modify the solution in this chapter to handle selection without replacement. Hint: add instance variables to Cookie to represent the hypothetical state of the bowls, and modify Likelihood accordingly. You might want to define a Bowl object.

CHAPTER 3
Estimation

The dice problem

Suppose I have a box of dice that contains a 4-sided die, a 6-sided die, an 8-sided die, a 12-sided die, and a 20-sided die. If you have ever played *Dungeons & Dragons*, you know what I am talking about.

Suppose I select a die from the box at random, roll it, and get a 6. What is the probability that I rolled each die?

Let me suggest a three-step strategy for approaching a problem like this.

1. Choose a representation for the hypotheses.
2. Choose a representation for the data.
3. Write the likelihood function.

In previous examples I used strings to represent hypotheses and data, but for the die problem I'll use numbers. Specifically, I'll use the integers 4, 6, 8, 12, and 20 to represent hypotheses:

```
suite = Dice([4, 6, 8, 12, 20])
```

And integers from 1 to 20 for the data. These representations make it easy to write the likelihood function:

```
class Dice(Suite):
    def Likelihood(self, data, hypo):
        if hypo < data:
            return 0
        else:
            return 1.0/hypo
```

Here's how `Likelihood` works. If `hypo<data`, that means the roll is greater than the number of sides on the die. That can't happen, so the likelihood is 0.

Otherwise the question is, "Given that there are `hypo` sides, what is the chance of rolling `data`?" The answer is `1/hypo`, regardless of `data`.

Here is the statement that does the update (if I roll a 6):

```
suite.Update(6)
```

And here is the posterior distribution:

```
4 0.0
6 0.392156862745
8 0.294117647059
12 0.196078431373
20 0.117647058824
```

After we roll a 6, the probability for the 4-sided die is 0. The most likely alternative is the 6-sided die, but there is still almost a 12% chance for the 20-sided die.

What if we roll a few more times and get 6, 8, 7, 7, 5, and 4?

```
for roll in [6, 8, 7, 7, 5, 4]:
    suite.Update(roll)
```

With this data the 6-sided die is eliminated, and the 8-sided die seems quite likely. Here are the results:

```
4 0.0
6 0.0
8 0.943248453672
12 0.0552061280613
20 0.0015454182665
```

Now the probability is 94% that we are rolling the 8-sided die, and less than 1% for the 20-sided die.

The dice problem is based on an example I saw in Sanjoy Mahajan's class on Bayesian inference. You can download the code in this section from *http://thinkbayes.com/dice.py*. For more information see "Working with the code" on page xi.

The locomotive problem

I found the locomotive problem in Frederick Mosteller's, *Fifty Challenging Problems in Probability with Solutions* (Dover, 1987):

> "A railroad numbers its locomotives in order 1..N. One day you see a locomotive with the number 60. Estimate how many locomotives the railroad has."

Based on this observation, we know the railroad has 60 or more locomotives. But how many more? To apply Bayesian reasoning, we can break this problem into two steps:

1. What did we know about N before we saw the data?
2. For any given value of N, what is the likelihood of seeing the data (a locomotive with number 60)?

The answer to the first question is the prior. The answer to the second is the likelihood.

We don't have much basis to choose a prior, but we can start with something simple and then consider alternatives. Let's assume that N is equally likely to be any value from 1 to 1000.

```
hypos = xrange(1, 1001)
```

Now all we need is a likelihood function. In a hypothetical fleet of N locomotives, what is the probability that we would see number 60? If we assume that there is only one train-operating company (or only one we care about) and that we are equally likely to see any of its locomotives, then the chance of seeing any particular locomotive is $1/N$.

Here's the likelihood function:

```
class Train(Suite):
    def Likelihood(self, data, hypo):
        if hypo < data:
            return 0
        else:
            return 1.0/hypo
```

This might look familiar; the likelihood functions for the locomotive problem and the dice problem are identical.

Here's the update:

```
suite = Train(hypos)
suite.Update(60)
```

There are too many hypotheses to print, so I plotted the results in Figure 3-1. Not surprisingly, all values of N below 60 have been eliminated.

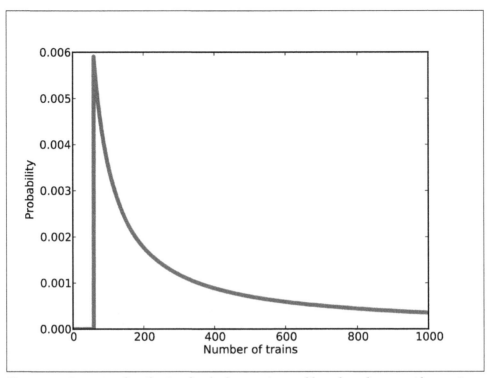

Figure 3-1. Posterior distribution for the locomotive problem, based on a uniform prior.

The most likely value, if you had to guess, is 60. That might not seem like a very good guess; after all, what are the chances that you just happened to see the train with the highest number? Nevertheless, if you want to maximize the chance of getting the answer exactly right, you should guess 60.

But maybe that's not the right goal. An alternative is to compute the mean of the posterior distribution:

```
def Mean(suite):
    total = 0
    for hypo, prob in suite.Items():
        total += hypo * prob
    return total

print Mean(suite)
```

Or you could use the very similar method provided by Pmf:

```
print suite.Mean()
```

The mean of the posterior is 333, so that might be a good guess if you wanted to minimize error. If you played this guessing game over and over, using the mean of the posterior as your estimate would minimize the mean squared error over the long run (see *http://en.wikipedia.org/wiki/Minimum_mean_square_error*).

You can download this example from *http://thinkbayes.com/train.py*. For more information see "Working with the code" on page xi.

What about that prior?

To make any progress on the locomotive problem we had to make assumptions, and some of them were pretty arbitrary. In particular, we chose a uniform prior from 1 to 1000, without much justification for choosing 1000, or for choosing a uniform distribution.

It is not crazy to believe that a railroad company might operate 1000 locomotives, but a reasonable person might guess more or fewer. So we might wonder whether the posterior distribution is sensitive to these assumptions. With so little data—only one observation—it probably is.

Recall that with a uniform prior from 1 to 1000, the mean of the posterior is 333. With an upper bound of 500, we get a posterior mean of 207, and with an upper bound of 2000, the posterior mean is 552.

So that's bad. There are two ways to proceed:

- Get more data.
- Get more background information.

With more data, posterior distributions based on different priors tend to converge. For example, suppose that in addition to train 60 we also see trains 30 and 90. We can update the distribution like this:

```
for data in [60, 30, 90]:
    suite.Update(data)
```

With these data, the means of the posteriors are

Upper Bound	Posterior Mean
500	152
1000	164
2000	171

So the differences are smaller.

An alternative prior

If more data are not available, another option is to improve the priors by gathering more background information. It is probably not reasonable to assume that a train-operating company with 1000 locomotives is just as likely as a company with only 1.

With some effort, we could probably find a list of companies that operate locomotives in the area of observation. Or we could interview an expert in rail shipping to gather information about the typical size of companies.

But even without getting into the specifics of railroad economics, we can make some educated guesses. In most fields, there are many small companies, fewer medium-sized companies, and only one or two very large companies. In fact, the distribution of company sizes tends to follow a power law, as Robert Axtell reports in *Science* (see *http://www.sciencemag.org/content/293/5536/1818.full.pdf*).

This law suggests that if there are 1000 companies with fewer than 10 locomotives, there might be 100 companies with 100 locomotives, 10 companies with 1000, and possibly one company with 10,000 locomotives.

Mathematically, a power law means that the number of companies with a given size is inversely proportional to size, or

$$\text{PMF}(x) \propto \left(\frac{1}{x}\right)^\alpha$$

where PMF(x) is the probability mass function of x and α is a parameter that is often near 1.

We can construct a power law prior like this:

```
class Train(Dice):

    def __init__(self, hypos, alpha=1.0):
        Pmf.__init__(self)
        for hypo in hypos:
            self.Set(hypo, hypo**(-alpha))
        self.Normalize()
```

And here's the code that constructs the prior:

```
hypos = range(1, 1001)
suite = Train(hypos)
```

Again, the upper bound is arbitrary, but with a power law prior, the posterior is less sensitive to this choice.

Figure 3-2 shows the new posterior based on the power law, compared to the posterior based on the uniform prior. Using the background information represented in the power law prior, we can all but eliminate values of N greater than 700.

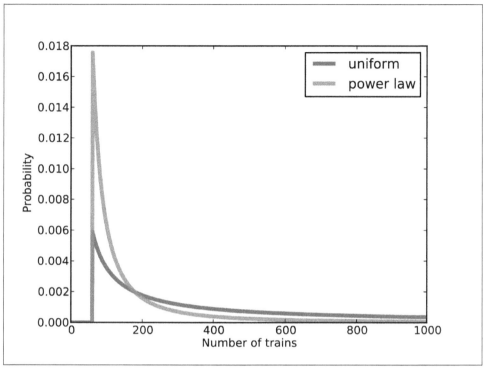

Figure 3-2. Posterior distribution based on a power law prior, compared to a uniform prior.

If we start with this prior and observe trains 30, 60, and 90, the means of the posteriors are:

Upper Bound	Posterior Mean
500	131
1000	133
2000	134

Now the differences are much smaller. In fact, with an arbitrarily large upper bound, the mean converges on 134.

So the power law prior is more realistic, because it is based on general information about the size of companies, and it behaves better in practice.

You can download the examples in this section from *http://thinkbayes.com/train3.py*. For more information see "Working with the code" on page xi.

Credible intervals

Once you have computed a posterior distribution, it is often useful to summarize the results with a single point estimate or an interval. For point estimates it is common to use the mean, median, or the value with maximum likelihood.

For intervals we usually report two values computed so that there is a 90% chance that the unknown value falls between them (or any other probability). These values define a **credible interval**.

A simple way to compute a credible interval is to add up the probabilities in the posterior distribution and record the values that correspond to probabilities 5% and 95%. In other words, the 5th and 95th percentiles.

thinkbayes provides a function that computes percentiles:

```
def Percentile(pmf, percentage):
    p = percentage / 100.0
    total = 0
    for val, prob in pmf.Items():
        total += prob
        if total >= p:
            return val
```

And here's the code that uses it:

```
interval = Percentile(suite, 5), Percentile(suite, 95)
print interval
```

For the previous example—the locomotive problem with a power law prior and three trains—the 90% credible interval is (91, 243). The width of this range suggests, correctly, that we are still quite uncertain about how many locomotives there are.

Cumulative distribution functions

In the previous section we computed percentiles by iterating through the values and probabilities in a Pmf. If we need to compute more than a few percentiles, it is more efficient to use a cumulative distribution function, or Cdf.

Cdfs and Pmfs are equivalent in the sense that they contain the same information about the distribution, and you can always convert from one to the other. The advantage of the Cdf is that you can compute percentiles more efficiently.

thinkbayes provides a Cdf class that represents a cumulative distribution function. Pmf provides a method that makes the corresponding Cdf:

```
cdf = suite.MakeCdf()
```

And `Cdf` provides a function named `Percentile`

```
interval = cdf.Percentile(5), cdf.Percentile(95)
```

Converting from a Pmf to a Cdf takes time proportional to the number of values, `len(pmf)`. The Cdf stores the values and probabilities in sorted lists, so looking up a probability to get the corresponding value takes "log time": that is, time proportional to the logarithm of the number of values. Looking up a value to get the corresponding probability is also logarithmic, so Cdfs are efficient for many calculations.

The examples in this section are in *http://thinkbayes.com/train3.py*. For more information see "Working with the code" on page xi.

The German tank problem

During World War II, the Economic Warfare Division of the American Embassy in London used statistical analysis to estimate German production of tanks and other equipment.[1]

The Western Allies had captured log books, inventories, and repair records that included chassis and engine serial numbers for individual tanks.

Analysis of these records indicated that serial numbers were allocated by manufacturer and tank type in blocks of 100 numbers, that numbers in each block were used sequentially, and that not all numbers in each block were used. So the problem of estimating German tank production could be reduced, within each block of 100 numbers, to a form of the locomotive problem.

Based on this insight, American and British analysts produced estimates substantially lower than estimates from other forms of intelligence. And after the war, records indicated that they were substantially more accurate.

They performed similar analyses for tires, trucks, rockets, and other equipment, yielding accurate and actionable economic intelligence.

The German tank problem is historically interesting; it is also a nice example of real-world application of statistical estimation. So far many of the examples in this book have been toy problems, but it will not be long before we start solving real problems. I think it is an advantage of Bayesian analysis, especially with the computational approach we are taking, that it provides such a short path from a basic introduction to the research frontier.

---

1 Ruggles and Brodie, "An Empirical Approach to Economic Intelligence in World War II," *Journal of the American Statistical Association*, Vol. 42, No. 237 (March 1947).

Discussion

Among Bayesians, there are two approaches to choosing prior distributions. Some recommend choosing the prior that best represents background information about the problem; in that case the prior is said to be **informative**. The problem with using an informative prior is that people might use different background information (or interpret it differently). So informative priors often seem subjective.

The alternative is a so-called **uninformative prior**, which is intended to be as unrestricted as possible, in order to let the data speak for themselves. In some cases you can identify a unique prior that has some desirable property, like representing minimal prior information about the estimated quantity.

Uninformative priors are appealing because they seem more objective. But I am generally in favor of using informative priors. Why? First, Bayesian analysis is always based on modeling decisions. Choosing the prior is one of those decisions, but it is not the only one, and it might not even be the most subjective. So even if an uninformative prior is more objective, the entire analysis is still subjective.

Also, for most practical problems, you are likely to be in one of two regimes: either you have a lot of data or not very much. If you have a lot of data, the choice of the prior doesn't matter very much; informative and uninformative priors yield almost the same results. We'll see an example like this in the next chapter.

But if, as in the locomotive problem, you don't have much data, using relevant background information (like the power law distribution) makes a big difference.

And if, as in the German tank problem, you have to make life-and-death decisions based on your results, you should probably use all of the information at your disposal, rather than maintaining the illusion of objectivity by pretending to know less than you do.

Exercises

Exercise 3-1.

To write a likelihood function for the locomotive problem, we had to answer this question: "If the railroad has N locomotives, what is the probability that we see number 60?"

The answer depends on what sampling process we use when we observe the locomotive. In this chapter, I resolved the ambiguity by specifying that there is only one train-operating company (or only one that we care about).

But suppose instead that there are many companies with different numbers of trains. And suppose that you are equally likely to see any train operated by any company. In that case, the likelihood function is different because you are more likely to see a train operated by a large company.

As an exercise, implement the likelihood function for this variation of the locomotive problem, and compare the results.

CHAPTER 4
More Estimation

The Euro problem

In *Information Theory, Inference, and Learning Algorithms*, David MacKay poses this problem:

> A statistical statement appeared in "The Guardian" on Friday January 4, 2002:
>
>> When spun on edge 250 times, a Belgian one-euro coin came up heads 140 times and tails 110. 'It looks very suspicious to me,' said Barry Blight, a statistics lecturer at the London School of Economics. 'If the coin were unbiased, the chance of getting a result as extreme as that would be less than 7%.'
>
> But do these data give evidence that the coin is biased rather than fair?

To answer that question, we'll proceed in two steps. The first is to estimate the probability that the coin lands face up. The second is to evaluate whether the data support the hypothesis that the coin is biased.

You can download the code in this section from *http://thinkbayes.com/euro.py*. For more information see "Working with the code" on page xi.

Any given coin has some probability, x, of landing heads up when spun on edge. It seems reasonable to believe that the value of x depends on some physical characteristics of the coin, primarily the distribution of weight.

If a coin is perfectly balanced, we expect x to be close to 50%, but for a lopsided coin, x might be substantially different. We can use Bayes's theorem and the observed data to estimate x.

Let's define 101 hypotheses, where H_x is the hypothesis that the probability of heads is x%, for values from 0 to 100. I'll start with a uniform prior where the probability of H_x is the same for all x. We'll come back later to consider other priors.

The likelihood function is relatively easy: If H_x is true, the probability of heads is $x/100$ and the probability of tails is $1 - x/100$.

```
class Euro(Suite):

    def Likelihood(self, data, hypo):
        x = hypo
        if data == 'H':
            return x/100.0
        else:
            return 1 - x/100.0
```

Here's the code that makes the suite and updates it:

```
suite = Euro(xrange(0, 101))
dataset = 'H' * 140 + 'T' * 110

for data in dataset:
    suite.Update(data)
```

The result is in Figure 4-1.

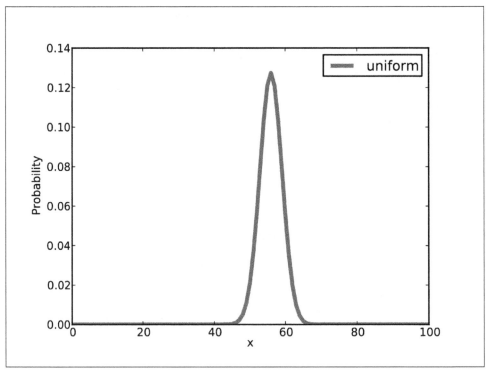

Figure 4-1. Posterior distribution for the Euro problem on a uniform prior.

Summarizing the posterior

Again, there are several ways to summarize the posterior distribution. One option is to find the most likely value in the posterior distribution. `thinkbayes` provides a function that does that:

```
def MaximumLikelihood(pmf):
    """Returns the value with the highest probability."""
    prob, val = max((prob, val) for val, prob in pmf.Items())
    return val
```

In this case the result is 56, which is also the observed percentage of heads, 140/250 = 56%. So that suggests (correctly) that the observed percentage is the maximum likelihood estimator for the population.

We might also summarize the posterior by computing the mean and median:

```
print 'Mean', suite.Mean()
print 'Median', thinkbayes.Percentile(suite, 50)
```

The mean is 55.95; the median is 56. Finally, we can compute a credible interval:

```
print 'CI', thinkbayes.CredibleInterval(suite, 90)
```

The result is (51, 61).

Now, getting back to the original question, we would like to know whether the coin is fair. We observe that the posterior credible interval does not include 50%, which suggests that the coin is not fair.

But that is not exactly the question we started with. MacKay asked, " Do these data give evidence that the coin is biased rather than fair?" To answer that question, we will have to be more precise about what it means to say that data constitute evidence for a hypothesis. And that is the subject of the next chapter.

But before we go on, I want to address one possible source of confusion. Since we want to know whether the coin is fair, it might be tempting to ask for the probability that x is 50%:

```
print suite.Prob(50)
```

The result is 0.021, but that value is almost meaningless. The decision to evaluate 101 hypotheses was arbitrary; we could have divided the range into more or fewer pieces, and if we had, the probability for any given hypothesis would be greater or less.

Swamping the priors

We started with a uniform prior, but that might not be a good choice. I can believe that if a coin is lopsided, x might deviate substantially from 50%, but it seems unlikely that the Belgian Euro coin is so imbalanced that x is 10% or 90%.

It might be more reasonable to choose a prior that gives higher probability to values of x near 50% and lower probability to extreme values.

As an example, I constructed a triangular prior, shown in Figure 4-2. Here's the code that constructs the prior:

```
def TrianglePrior():
    suite = Euro()
    for x in range(0, 51):
        suite.Set(x, x)
    for x in range(51, 101):
        suite.Set(x, 100-x)
    suite.Normalize()
```

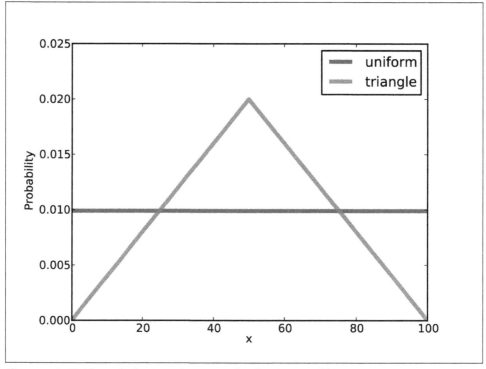

Figure 4-2. Uniform and triangular priors for the Euro problem.

Figure 4-2 shows the result (and the uniform prior for comparison). Updating this prior with the same dataset yields the posterior distribution shown in Figure 4-3. Even with substantially different priors, the posterior distributions are very similar. The medians and the credible intervals are identical; the means differ by less than 0.5%.

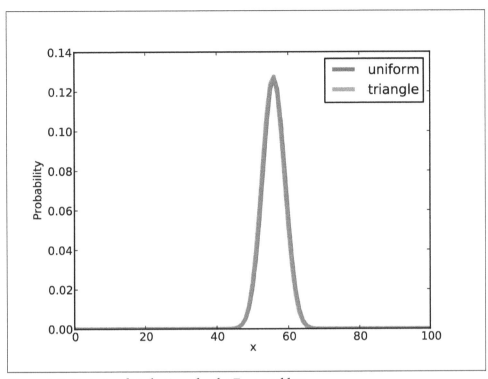

Figure 4-3. Posterior distributions for the Euro problem.

This is an example of **swamping the priors**: with enough data, people who start with different priors will tend to converge on the same posterior.

Optimization

The code I have shown so far is meant to be easy to read, but it is not very efficient. In general, I like to develop code that is demonstrably correct, then check whether it is fast enough for my purposes. If so, there is no need to optimize. For this example, if we care about run time, there are several ways we can speed it up.

The first opportunity is to reduce the number of times we normalize the suite. In the original code, we call `Update` once for each spin.

```
dataset = 'H' * heads + 'T' * tails

for data in dataset:
    suite.Update(data)
```

And here's what `Update` looks like:

```
def Update(self, data):
    for hypo in self.Values():
        like = self.Likelihood(data, hypo)
        self.Mult(hypo, like)
    return self.Normalize()
```

Each update iterates through the hypotheses, then calls `Normalize`, which iterates through the hypotheses again. We can save some time by doing all of the updates before normalizing.

`Suite` provides a method called `UpdateSet` that does exactly that. Here it is:

```
def UpdateSet(self, dataset):
    for data in dataset:
        for hypo in self.Values():
            like = self.Likelihood(data, hypo)
            self.Mult(hypo, like)
    return self.Normalize()
```

And here's how we can invoke it:

```
dataset = 'H' * heads + 'T' * tails
suite.UpdateSet(dataset)
```

This optimization speeds things up, but the run time is still proportional to the amount of data. We can speed things up even more by rewriting `Likelihood` to process the entire dataset, rather than one spin at a time.

In the original version, `data` is a string that encodes either heads or tails:

```
def Likelihood(self, data, hypo):
    x = hypo / 100.0
    if data == 'H':
        return x
    else:
        return 1-x
```

As an alternative, we could encode the dataset as a tuple of two integers: the number of heads and tails. In that case `Likelihood` looks like this:

```
def Likelihood(self, data, hypo):
    x = hypo / 100.0
    heads, tails = data
    like = x**heads * (1-x)**tails
    return like
```

And then we can call `Update` like this:

```
heads, tails = 140, 110
suite.Update((heads, tails))
```

Since we have replaced repeated multiplication with exponentiation, this version takes the same time for any number of spins.

The beta distribution

There is one more optimization that solves this problem even faster.

So far we have used a Pmf object to represent a discrete set of values for x. Now we will use a continuous distribution, specifically the beta distribution (see *http://en.wikipedia.org/wiki/Beta_distribution*).

The beta distribution is defined on the interval from 0 to 1 (including both), so it is a natural choice for describing proportions and probabilities. But wait, it gets better.

It turns out that if you do a Bayesian update with a binomial likelihood function, which is what we did in the previous section, the beta distribution is a **conjugate prior**. That means that if the prior distribution for x is a beta distribution, the posterior is also a beta distribution. But wait, it gets even better.

The shape of the beta distribution depends on two parameters, written α and β, or `alpha` and `beta`. If the prior is a beta distribution with parameters `alpha` and `beta`, and we see data with h heads and t tails, the posterior is a beta distribution with parameters `alpha+h` and `beta+t`. In other words, we can do an update with two additions.

So that's great, but it only works if we can find a beta distribution that is a good choice for a prior. Fortunately, for many realistic priors there is a beta distribution that is at least a good approximation, and for a uniform prior there is a perfect match. The beta distribution with `alpha=1` and `beta=1` is uniform from 0 to 1.

Let's see how we can take advantage of all this. `thinkbayes.py` provides a class that represents a beta distribution:

```
class Beta(object):

    def __init__(self, alpha=1, beta=1):
        self.alpha = alpha
        self.beta = beta
```

By default `__init__` makes a uniform distribution. `Update` performs a Bayesian update:

```
def Update(self, data):
    heads, tails = data
    self.alpha += heads
    self.beta += tails
```

`data` is a pair of integers representing the number of heads and tails.

So we have yet another way to solve the Euro problem:

```
beta = thinkbayes.Beta()
beta.Update((140, 110))
print beta.Mean()
```

Beta provides Mean, which computes a simple function of alpha and beta:

```
def Mean(self):
    return float(self.alpha) / (self.alpha + self.beta)
```

For the Euro problem the posterior mean is 56%, which is the same result we got using Pmfs.

Beta also provides EvalPdf, which evaluates the probability density function (PDF) of the beta distribution:

```
def EvalPdf(self, x):
    return x**(self.alpha-1) * (1-x)**(self.beta-1)
```

Finally, Beta provides MakePmf, which uses EvalPdf to generate a discrete approximation of the beta distribution.

Discussion

In this chapter we solved the same problem with two different priors and found that with a large dataset, the priors get swamped. If two people start with different prior beliefs, they generally find, as they see more data, that their posterior distributions converge. At some point the difference between their distributions is small enough that it has no practical effect.

When this happens, it relieves some of the worry about objectivity that I discussed in the previous chapter. And for many real-world problems even stark prior beliefs can eventually be reconciled by data.

But that is not always the case. First, remember that all Bayesian analysis is based on modeling decisions. If you and I do not choose the same model, we might interpret data differently. So even with the same data, we would compute different likelihoods, and our posterior beliefs might not converge.

Also, notice that in a Bayesian update, we multiply each prior probability by a likelihood, so if $p(H)$ is 0, $p(H|D)$ is also 0, regardless of D. In the Euro problem, if you are convinced that x is less than 50%, and you assign probability 0 to all other hypotheses, no amount of data will convince you otherwise.

This observation is the basis of **Cromwell's rule**, which is the recommendation that you should avoid giving a prior probability of 0 to any hypothesis that is even remotely possible (see *http://en.wikipedia.org/wiki/Cromwell's_rule*).

Cromwell's rule is named after Oliver Cromwell, who wrote, "I beseech you, in the bowels of Christ, think it possible that you may be mistaken." For Bayesians, this turns out to be good advice (even if it's a little overwrought).

Exercises

Exercise 4-1.

Suppose that instead of observing coin tosses directly, you measure the outcome using an instrument that is not always correct. Specifically, suppose there is a probability y that an actual heads is reported as tails, or actual tails reported as heads.

Write a class that estimates the bias of a coin given a series of outcomes and the value of y.

How does the spread of the posterior distribution depend on y?

Exercise 4-2.

This exercise is inspired by a question posted by a "redditor" named dominosci on Reddit's statistics "subreddit" at *http://reddit.com/r/statistics*.

Reddit is an online forum with many interest groups called subreddits. Users, called redditors, post links to online content and other web pages. Other redditors vote on the links, giving an "upvote" to high-quality links and a "downvote" to links that are bad or irrelevant.

A problem, identified by dominosci, is that some redditors are more reliable than others, and Reddit does not take this into account.

The challenge is to devise a system so that when a redditor casts a vote, the estimated quality of the link is updated in accordance with the reliability of the redditor, and the estimated reliability of the redditor is updated in accordance with the quality of the link.

One approach is to model the quality of the link as the probability of garnering an upvote, and to model the reliability of the redditor as the probability of correctly giving an upvote to a high-quality item.

Write class definitions for redditors and links and an update function that updates both objects whenever a redditor casts a vote.

CHAPTER 5
Odds and Addends

Odds

One way to represent a probability is with a number between 0 and 1, but that's not the only way. If you have ever bet on a football game or a horse race, you have probably encountered another representation of probability, called **odds**.

You might have heard expressions like "the odds are three to one," but you might not know what they mean. The **odds in favor** of an event are the ratio of the probability it will occur to the probability that it will not.

So if I think my team has a 75% chance of winning, I would say that the odds in their favor are three to one, because the chance of winning is three times the chance of losing.

You can write odds in decimal form, but it is most common to write them as a ratio of integers. So "three to one" is written 3:1.

When probabilities are low, it is more common to report the **odds against** rather than the odds in favor. For example, if I think my horse has a 10% chance of winning, I would say that the odds against are 9:1.

Probabilities and odds are different representations of the same information. Given a probability, you can compute the odds like this:

```
def Odds(p):
    return p / (1-p)
```

Given the odds in favor, in decimal form, you can convert to probability like this:

```
def Probability(o):
    return o / (o+1)
```

If you represent odds with a numerator and denominator, you can convert to probability like this:

```
def Probability2(yes, no):
    return yes / (yes + no)
```

When I work with odds in my head, I find it helpful to picture people at the track. If 20% of them think my horse will win, then 80% of them don't, so the odds in favor are 20:80 or 1:4.

If the odds are 5:1 against my horse, then five out of six people think she will lose, so the probability of winning is 1/6.

The odds form of Bayes's theorem

In Chapter 1 I wrote Bayes's theorem in the **probability form**:

$$p(H|D) = \frac{p(H)\, p(D|H)}{p(D)}$$

If we have two hypotheses, *A* and *B*, we can write the ratio of posterior probabilities like this:

$$\frac{p(A|D)}{p(B|D)} = \frac{p(A)\, p(D|A)}{p(B)\, p(D|B)}$$

Notice that the normalizing constant, p(*D*), drops out of this equation.

If *A* and *B* are mutually exclusive and collectively exhaustive, that means p(*B*) = 1 − p(*A*), so we can rewrite the ratio of the priors, and the ratio of the posteriors, as odds.

Writing o(*A*) for odds in favor of *A*, we get:

$$o(A|D) = o(A)\, \frac{p(D|A)}{p(D|B)}$$

In words, this says that the posterior odds are the prior odds times the likelihood ratio. This is the **odds form** of Bayes's theorem.

This form is most convenient for computing a Bayesian update on paper or in your head. For example, let's go back to the cookie problem:

Suppose there are two bowls of cookies. Bowl 1 contains 30 vanilla cookies and 10 chocolate cookies. Bowl 2 contains 20 of each.

Now suppose you choose one of the bowls at random and, without looking, select a cookie at random. The cookie is vanilla. What is the probability that it came from Bowl 1?

The prior probability is 50%, so the prior odds are 1:1, or just 1. The likelihood ratio is $\frac{3}{4}/\frac{1}{2}$, or 3/2. So the posterior odds are 3:2, which corresponds to probability 3/5.

Oliver's blood

Here is another problem from MacKay's *Information Theory, Inference, and Learning Algorithms*:

Two people have left traces of their own blood at the scene of a crime. A suspect, Oliver, is tested and found to have type 'O' blood. The blood groups of the two traces are found to be of type 'O' (a common type in the local population, having frequency 60%) and of type 'AB' (a rare type, with frequency 1%). Do these data [the traces found at the scene] give evidence in favor of the proposition that Oliver was one of the people [who left blood at the scene]?

To answer this question, we need to think about what it means for data to give evidence in favor of (or against) a hypothesis. Intuitively, we might say that data favor a hypothesis if the hypothesis is more likely in light of the data than it was before.

In the cookie problem, the prior odds are 1:1, or probability 50%. The posterior odds are 3:2, or probability 60%. So we could say that the vanilla cookie is evidence in favor of Bowl 1.

The odds form of Bayes's theorem provides a way to make this intuition more precise. Again

$$o(A|D) = o(A) \frac{p(D|A)}{p(D|B)}$$

Or dividing through by o(A):

$$\frac{o(A|D)}{o(A)} = \frac{p(D|A)}{p(D|B)}$$

The term on the left is the ratio of the posterior and prior odds. The term on the right is the likelihood ratio, also called the **Bayes factor**.

If the Bayes factor value is greater than 1, that means that the data were more likely under *A* than under *B*. And since the odds ratio is also greater than 1, that means that the odds are greater, in light of the data, than they were before.

If the Bayes factor is less than 1, that means the data were less likely under *A* than under *B*, so the odds in favor of *A* go down.

Finally, if the Bayes factor is exactly 1, the data are equally likely under either hypothesis, so the odds do not change.

Now we can get back to the Oliver's blood problem. If Oliver is one of the people who left blood at the crime scene, then he accounts for the 'O' sample, so the probability of the data is just the probability that a random member of the population has type 'AB' blood, which is 1%.

If Oliver did not leave blood at the scene, then we have two samples to account for. If we choose two random people from the population, what is the chance of finding one with type 'O' and one with type 'AB'? Well, there are two ways it might happen: the first person we choose might have type 'O' and the second 'AB', or the other way around. So the total probability is $2(0.6)(0.01) = 1.2\%$.

The likelihood of the data is slightly higher if Oliver is *not* one of the people who left blood at the scene, so the blood data is actually evidence against Oliver's guilt.

This example is a little contrived, but it is an example of the counterintuitive result that data *consistent* with a hypothesis are not necessarily *in favor of* the hypothesis.

If this result is so counterintuitive that it bothers you, this way of thinking might help: the data consist of a common event, type 'O' blood, and a rare event, type 'AB' blood. If Oliver accounts for the common event, that leaves the rare event still unexplained. If Oliver doesn't account for the 'O' blood, then we have two chances to find someone in the population with 'AB' blood. And that factor of two makes the difference.

Addends

The fundamental operation of Bayesian statistics is `Update`, which takes a prior distribution and a set of data, and produces a posterior distribution. But solving real problems usually involves a number of other operations, including scaling, addition and other arithmetic operations, max and min, and mixtures.

This chapter presents addition and max; I will present other operations as we need them.

The first example is based on *Dungeons & Dragons*, a role-playing game where the results of players' decisions are usually determined by rolling dice. In fact, before game play starts, players generate each attribute of their characters—strength, intelligence, wisdom, dexterity, constitution, and charisma—by rolling three 6-sided dice and adding them up.

So you might be curious to know the distribution of this sum. There are two ways you might compute it:

Simulation:
> Given a Pmf that represents the distribution for a single die, you can draw random samples, add them up, and accumulate the distribution of simulated sums.

Enumeration:
> Given two Pmfs, you can enumerate all possible pairs of values and compute the distribution of the sums.

thinkbayes provides functions for both. Here's an example of the first approach. First, I'll define a class to represent a single die as a Pmf:

```
class Die(thinkbayes.Pmf):

    def __init__(self, sides):
        thinkbayes.Pmf.__init__(self)
        for x in xrange(1, sides+1):
            self.Set(x, 1)
        self.Normalize()
```

Now I can create a 6-sided die:

```
d6 = Die(6)
```

And use thinkbayes.SampleSum to generate a sample of 1000 rolls.

```
dice = [d6] * 3
three = thinkbayes.SampleSum(dice, 1000)
```

SampleSum takes list of distributions (either Pmf or Cdf objects) and the sample size, n. It generates n random sums and returns their distribution as a Pmf object.

```
def SampleSum(dists, n):
    pmf = MakePmfFromList(RandomSum(dists) for i in xrange(n))
    return pmf
```

SampleSum uses RandomSum, also in thinkbayes.py:

```
def RandomSum(dists):
    total = sum(dist.Random() for dist in dists)
    return total
```

RandomSum invokes Random on each distribution and adds up the results.

The drawback of simulation is that the result is only approximately correct. As n gets larger, it gets more accurate, but of course the run time increases as well.

The other approach is to enumerate all pairs of values and compute the sum and probability of each pair. This is implemented in Pmf.__add__:

```
# class Pmf

    def __add__(self, other):
        pmf = Pmf()
        for v1, p1 in self.Items():
            for v2, p2 in other.Items():
                pmf.Incr(v1+v2, p1*p2)
        return pmf
```

self is a Pmf, of course; other can be a Pmf or anything else that provides Items. The result is a new Pmf. The time to run __add__ depends on the number of items in self and other; it is proportional to len(self) * len(other).

And here's how it's used:

```
three_exact = d6 + d6 + d6
```

When you apply the + operator to a Pmf, Python invokes __add__. In this example, __add__ is invoked twice.

Figure 5-1 shows an approximate result generated by simulation and the exact result computed by enumeration.

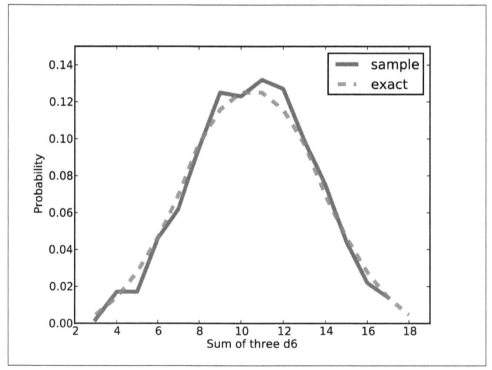

Figure 5-1. Approximate and exact distributions for the sum of three 6-sided dice.

Pmf.__add__ is based on the assumption that the random selections from each Pmf are independent. In the example of rolling several dice, this assumption is pretty good. In other cases, we would have to extend this method to use conditional probabilities.

The code from this section is available from *http://thinkbayes.com/dungeons.py*. For more information see "Working with the code" on page xi.

Maxima

When you generate a *Dungeons & Dragons* character, you are particularly interested in the character's best attributes, so you might like to know the distribution of the maximum attribute.

There are three ways to compute the distribution of a maximum:

Simulation:
    Given a Pmf that represents the distribution for a single selection, you can generate random samples, find the maximum, and accumulate the distribution of simulated maxima.

Enumeration:
    Given two Pmfs, you can enumerate all possible pairs of values and compute the distribution of the maximum.

Exponentiation:
    If we convert a Pmf to a Cdf, there is a simple and efficient algorithm for finding the Cdf of the maximum.

The code to simulate maxima is almost identical to the code for simulating sums:

```
def RandomMax(dists):
    total = max(dist.Random() for dist in dists)
    return total

def SampleMax(dists, n):
    pmf = MakePmfFromList(RandomMax(dists) for i in xrange(n))
    return pmf
```

All I did was replace "sum" with "max". And the code for enumeration is almost identical, too:

```
def PmfMax(pmf1, pmf2):
    res = thinkbayes.Pmf()
    for v1, p1 in pmf1.Items():
        for v2, p2 in pmf2.Items():
            res.Incr(max(v1, v2), p1*p2)
    return res
```

In fact, you could generalize this function by taking the appropriate operator as a parameter.

The only problem with this algorithm is that if each Pmf has m values, the run time is proportional to m^2. And if we want the maximum of k selections, it takes time proportional to km^2.

If we convert the Pmfs to Cdfs, we can do the same calculation much faster! The key is to remember the definition of the cumulative distribution function:

$$CDF(x) = p(X \le x)$$

where X is a random variable that means "a value chosen randomly from this distribution." So, for example, $CDF(5)$ is the probability that a value from this distribution is less than or equal to 5.

If I draw X from CDF_1 and Y from CDF_2, and compute the maximum $Z = max(X, Y)$, what is the chance that Z is less than or equal to 5? Well, in that case both X and Y must be less than or equal to 5.

If the selections of X and Y are independent,

$$CDF_3(5) = CDF_1(5)CDF_2(5)$$

where CDF_3 is the distribution of Z. I chose the value 5 because I think it makes the formulas easy to read, but we can generalize for any value of z:

$$CDF_3(z) = CDF_1(z)CDF_2(z)$$

In the special case where we draw k values from the same distribution,

$$CDF_k(z) = CDF_1(z)^k$$

So to find the distribution of the maximum of k values, we can enumerate the probabilities in the given Cdf and raise them to the kth power. Cdf provides a method that does just that:

```
# class Cdf

    def Max(self, k):
        cdf = self.Copy()
        cdf.ps = [p**k for p in cdf.ps]
        return cdf
```

Max takes the number of selections, k, and returns a new Cdf that represents the distribution of the maximum of k selections. The run time for this method is proportional to *m*, the number of items in the Cdf.

Pmf.Max does the same thing for Pmfs. It has to do a little more work to convert the Pmf to a Cdf, so the run time is proportional to $m \log m$, but that's still better than quadratic.

Finally, here's an example that computes the distribution of a character's best attribute:

```
best_attr_cdf = three_exact.Max(6)
best_attr_pmf = best_attr_cdf.MakePmf()
```

Where three_exact is defined in the previous section. If we print the results, we see that the chance of generating a character with at least one attribute of 18 is about 3%. Figure 5-2 shows the distribution.

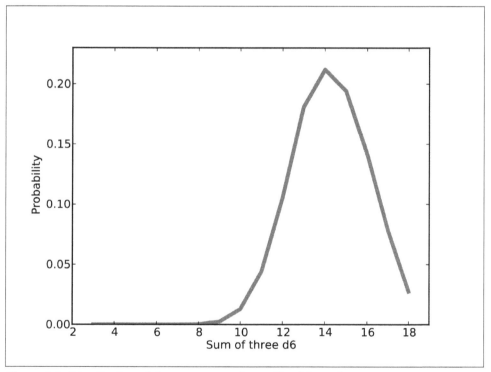

Figure 5-2. Distribution of the maximum of six rolls of three dice.

Mixtures

Let's do one more example from *Dungeons & Dragons*. Suppose I have a box of dice with the following inventory:

```
5  4-sided dice
4  6-sided dice
3  8-sided dice
2  12-sided dice
1  20-sided die
```

I choose a die from the box and roll it. What is the distribution of the outcome?

If you know which die it is, the answer is easy. A die with n sides yields a uniform distribution from 1 to n, including both.

But if we don't know which die it is, the resulting distribution is a **mixture** of uniform distributions with different bounds. In general, this kind of mixture does not fit any simple mathematical model, but it is straightforward to compute the distribution in the form of a PMF.

As always, one option is to simulate the scenario, generate a random sample, and compute the PMF of the sample. This approach is simple and it generates an approximate solution quickly. But if we want an exact solution, we need a different approach.

Let's start with a simple version of the problem where there are only two dice, one with 6 sides and one with 8. We can make a Pmf to represent each die:

```
d6 = Die(6)
d8 = Die(8)
```

Then we create a Pmf to represent the mixture:

```
mix = thinkbayes.Pmf()
for die in [d6, d8]:
    for outcome, prob in die.Items():
        mix.Incr(outcome, prob)
mix.Normalize()
```

The first loop enumerates the dice; the second enumerates the outcomes and their probabilities. Inside the loop, `Pmf.Incr` adds up the contributions from the two distributions.

This code assumes that the two dice are equally likely. More generally, we need to know the probability of each die so we can weight the outcomes accordingly.

First we create a Pmf that maps from each die to the probability it is selected:

```
pmf_dice = thinkbayes.Pmf()
pmf_dice.Set(Die(4), 5)
pmf_dice.Set(Die(6), 4)
pmf_dice.Set(Die(8), 3)
```

```
pmf_dice.Set(Die(12), 2)
pmf_dice.Set(Die(20), 1)
pmf_dice.Normalize()
```

Next we need a more general version of the mixture algorithm:

```
mix = thinkbayes.Pmf()
for die, weight in pmf_dice.Items():
    for outcome, prob in die.Items():
        mix.Incr(outcome, weight*prob)
```

Now each die has a weight associated with it (which makes it a weighted die, I suppose). When we add each outcome to the mixture, its probability is multiplied by weight.

Figure 5-3 shows the result. As expected, values 1 through 4 are the most likely because any die can produce them. Values above 12 are unlikely because there is only one die in the box that can produce them (and it does so less than half the time).

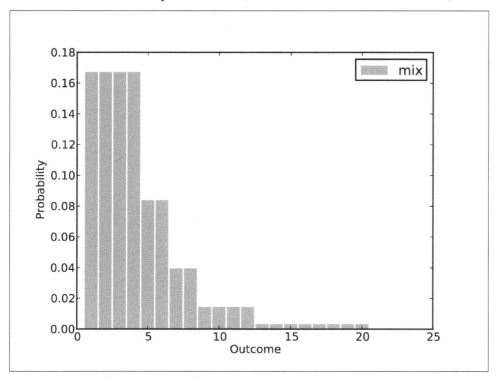

Figure 5-3. Distribution outcome for random die from a box.

thinkbayes provides a function named MakeMixture that encapsulates this algorithm, so we could have written:

```
mix = thinkbayes.MakeMixture(pmf_dice)
```

We'll use `MakeMixture` again in Chapters 7 and 8.

Discussion

Other than the odds form of Bayes's theorem, this chapter is not specifically Bayesian. But Bayesian analysis is all about distributions, so it is important to understand the concept of a distribution well. From a computational point of view, a distribution is any data structure that represents a set of values (possible outcomes of a random process) and their probabilities.

We have seen two representations of distributions: Pmfs and Cdfs. These representations are equivalent in the sense that they contain the same information, so you can convert from one to the other. The primary difference between them is performance: some operations are faster and easier with a Pmf; others are faster with a Cdf.

The other goal of this chapter is to introduce operations that act on distributions, like `Pmf.__add__`, `Cdf.Max`, and `thinkbayes.MakeMixture`. We will use these operations later, but I introduce them now to encourage you to think of a distribution as a fundamental unit of computation, not just a container for values and probabilities.

CHAPTER 6
Decision Analysis

The Price is Right problem

On November 1, 2007, contestants named Letia and Nathaniel appeared on *The Price is Right*, an American game show. They competed in a game called *The Showcase*, where the objective is to guess the price of a showcase of prizes. The contestant who comes closest to the actual price of the showcase, without going over, wins the prizes.

Nathaniel went first. His showcase included a dishwasher, a wine cabinet, a laptop computer, and a car. He bid $26,000.

Letia's showcase included a pinball machine, a video arcade game, a pool table, and a cruise of the Bahamas. She bid $21,500.

The actual price of Nathaniel's showcase was $25,347. His bid was too high, so he lost.

The actual price of Letia's showcase was $21,578. She was only off by $78, so she won her showcase and, because her bid was off by less than $250, she also won Nathaniel's showcase.

For a Bayesian thinker, this scenario suggests several questions:

1. Before seeing the prizes, what prior beliefs should the contestant have about the price of the showcase?
2. After seeing the prizes, how should the contestant update those beliefs?
3. Based on the posterior distribution, what should the contestant bid?

The third question demonstrates a common use of Bayesian analysis: decision analysis. Given a posterior distribution, we can choose the bid that maximizes the contestant's expected return.

This problem is inspired by an example in Cameron Davidson-Pilon's book, *Bayesian Methods for Hackers*. The code I wrote for this chapter is available from *http://thinkbayes.com/price.py*; it reads data files you can download from *http://thinkbayes.com/showcases.2011.csv* and *http://thinkbayes.com/showcases.2012.csv*. For more information see "Working with the code" on page xi.

The prior

To choose a prior distribution of prices, we can take advantage of data from previous episodes. Fortunately, fans of the show keep detailed records. When I corresponded with Mr. Davidson-Pilon about his book, he sent me data collected by Steve Gee at *http://tpirsummaries.8m.com*. It includes the price of each showcase from the 2011 and 2012 seasons and the bids offered by the contestants.

Figure 6-1 shows the distribution of prices for these showcases. The most common value for both showcases is around $28,000, but the first showcase has a second mode near $50,000, and the second showcase is occasionally worth more than $70,000.

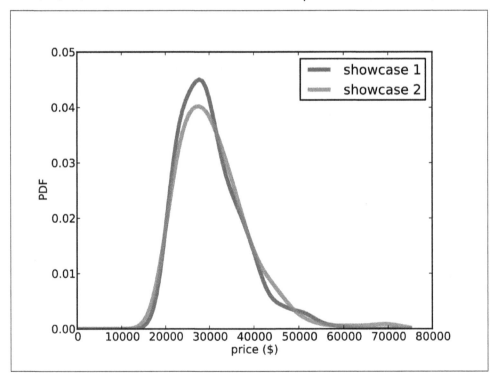

Figure 6-1. Distribution of prices for showcases on The Price is Right, 2011-12.

These distributions are based on actual data, but they have been smoothed by Gaussian kernel density estimation (KDE). Before we go on, I want to take a detour to talk about probability density functions and KDE.

Probability density functions

So far we have been working with probability mass functions, or PMFs. A PMF is a map from each possible value to its probability. In my implementation, a Pmf object provides a method named `Prob` that takes a value and returns a probability, also known as a **probability mass**.

In mathematical notation, PDFs are usually written as functions; for example, here is the PDF of a Gaussian distribution with mean 0 and standard deviation 1:

$$f(x) = \frac{1}{\sqrt{2\pi}} \exp\left(-x^2/2\right)$$

For a given value of x, this function computes a probability density. A density is similar to a probability mass in the sense that a higher density indicates that a value is more likely.

But a density is not a probability. A density can be 0 or any positive value; it is not bounded, like a probability, between 0 and 1.

If you integrate a density over a continuous range, the result is a probability. But for the applications in this book we seldom have to do that.

Instead we primarily use probability densities as part of a likelihood function. We will see an example soon.

Representing PDFs

To represent PDFs in Python, `thinkbayes.py` provides a class named `Pdf`. `Pdf` is an **abstract type**, which means that it defines the interface a Pdf is supposed to have, but does not provide a complete implementation. The `Pdf` interface includes two methods, `Density` and `MakePmf`:

```
class Pdf(object):

    def Density(self, x):
        raise UnimplementedMethodException()

    def MakePmf(self, xs):
        pmf = Pmf()
        for x in xs:
            pmf.Set(x, self.Density(x))
        pmf.Normalize()
        return pmf
```

`Density` takes a value, x, and returns the corresponding density. `MakePmf` makes a discrete approximation to the PDF.

`Pdf` provides an implementation of `MakePmf`, but not `Density`, which has to be provided by a child class.

A **concrete type** is a child class that extends an abstract type and provides an implementation of the missing methods. For example, `GaussianPdf` extends `Pdf` and provides `Density`:

```
class GaussianPdf(Pdf):

    def __init__(self, mu, sigma):
        self.mu = mu
        self.sigma = sigma

    def Density(self, x):
        return scipy.stats.norm.pdf(x, self.mu, self.sigma)
```

`__init__` takes `mu` and `sigma`, which are the mean and standard deviation of the distribution, and stores them as attributes.

`Density` uses a function from `scipy.stats` to evaluate the Gaussian PDF. The function is called `norm.pdf` because the Gaussian distribution is also called the "normal" distribution.

The Gaussian PDF is defined by a simple mathematical function, so it is easy to evaluate. And it is useful because many quantities in the real world have distributions that are approximately Gaussian.

But with real data, there is no guarantee that the distribution is Gaussian or any other simple mathematical function. In that case we can use a sample to estimate the PDF of the whole population.

For example, in *The Price Is Right* data, we have 313 prices for the first showcase. We can think of these values as a sample from the population of all possible showcase prices.

This sample includes the following values (in order):

28800, 28868, 28941, 28957, 28958

In the sample, no values appear between 28801 and 28867, but there is no reason to think that these values are impossible. Based on our background information, we expect all values in this range to be equally likely. In other words, we expect the PDF to be fairly smooth.

Kernel density estimation (KDE) is an algorithm that takes a sample and finds an appropriately smooth PDF that fits the data. You can read details at *http://en.wikipedia.org/wiki/Kernel_density_estimation*.

`scipy` provides an implementation of KDE and `thinkbayes` provides a class called `EstimatedPdf` that uses it:

```
class EstimatedPdf(Pdf):

    def __init__(self, sample):
        self.kde = scipy.stats.gaussian_kde(sample)

    def Density(self, x):
        return self.kde.evaluate(x)
```

`__init__` takes a sample and computes a kernel density estimate. The result is a `gaussian_kde` object that provides an `evaluate` method.

`Density` takes a value, calls `gaussian_kde.evaluate`, and returns the resulting density.

Finally, here's an outline of the code I used to generate Figure 6-1:

```
prices = ReadData()
pdf = thinkbayes.EstimatedPdf(prices)

low, high = 0, 75000
n = 101
xs = numpy.linspace(low, high, n)
pmf = pdf.MakePmf(xs)
```

`pdf` is a `Pdf` object, estimated by KDE. `pmf` is a `Pmf` object that approximates the `Pdf` by evaluating the density at a sequence of equally spaced values.

`linspace` stands for "linear space." It takes a range, `low` and `high`, and the number of points, `n`, and returns a new `numpy` array with n elements equally spaced between `low` and `high`, including both.

And now back to *The Price is Right*.

Modeling the contestants

The PDFs in Figure 6-1 estimate the distribution of possible prices. If you were a contestant on the show, you could use this distribution to quantify your prior belief about the price of each showcase (before you see the prizes).

To update these priors, we have to answer these questions:

1. What data should we consider and how should we quantify it?
2. Can we compute a likelihood function; that is, for each hypothetical value of `price`, can we compute the conditional likelihood of the data?

To answer these questions, I am going to model the contestant as a price-guessing instrument with known error characteristics. In other words, when the contestant sees the prizes, he or she guesses the price of each prize—ideally without taking into consideration the fact that the prize is part of a showcase—and adds up the prices. Let's call this total `guess`.

Under this model, the question we have to answer is, "If the actual price is `price`, what is the likelihood that the contestant's estimate would be `guess`?"

Or if we define:

```
error = price - guess
```

then we could ask, "What is the likelihood that the contestant's estimate is off by `error`?"

To answer this question, we can use the historical data again. Figure 6-2 shows the cumulative distribution of `diff`, the difference between the contestant's bid and the actual price of the showcase.

The definition of diff is:

```
diff = price - bid
```

When `diff` is negative, the bid is too high. As an aside, we can use this distribution to compute the probability that the contestants overbid: the first contestant overbids 25% of the time; the second contestant overbids 29% of the time.

We can also see that the bids are biased; that is, they are more likely to be too low than too high. And that makes sense, given the rules of the game.

Finally, we can use this distribution to estimate the reliability of the contestants' guesses. This step is a little tricky because we don't actually know the contestant's guesses; we only know what they bid.

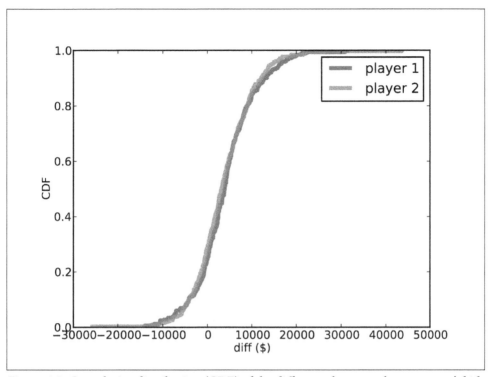

Figure 6-2. Cumulative distribution (CDF) of the difference between the contestant's bid and the actual price.

So we'll have to make some assumptions. Specifically, I assume that the distribution of error is Gaussian with mean 0 and the same variance as diff.

The Player class implements this model:

```
class Player(object):

    def __init__(self, prices, bids, diffs):
        self.pdf_price = thinkbayes.EstimatedPdf(prices)
        self.cdf_diff = thinkbayes.MakeCdfFromList(diffs)

        mu = 0
        sigma = numpy.std(diffs)
        self.pdf_error = thinkbayes.GaussianPdf(mu, sigma)
```

prices is a sequence of showcase prices, bids is a sequence of bids, and diffs is a sequence of diffs, where again diff = price - bid.

pdf_price is the smoothed PDF of prices, estimated by KDE. cdf_diff is the cumulative distribution of diff, which we saw in Figure 6-2. And pdf_error is the PDF that characterizes the distribution of errors; where error = price - guess.

Again, we use the variance of `diff` to estimate the variance of `error`. This estimate is not perfect because contestants' bids are sometimes strategic; for example, if Player 2 thinks that Player 1 has overbid, Player 2 might make a very low bid. In that case `diff` does not reflect `error`. If this happens a lot, the observed variance in `diff` might overestimate the variance in `error`. Nevertheless, I think it is a reasonable modeling decision.

As an alternative, someone preparing to appear on the show could estimate their own distribution of `error` by watching previous shows and recording their guesses and the actual prices.

Likelihood

Now we are ready to write the likelihood function. As usual, I define a new class that extends `thinkbayes.Suite`:

```
class Price(thinkbayes.Suite):

    def __init__(self, pmf, player):
        thinkbayes.Suite.__init__(self, pmf)
        self.player = player
```

`pmf` represents the prior distribution and `player` is a Player object as described in the previous section. Here's `Likelihood`:

```
    def Likelihood(self, data, hypo):
        price = hypo
        guess = data

        error = price - guess
        like = self.player.ErrorDensity(error)

        return like
```

`hypo` is the hypothetical price of the showcase. `data` is the contestant's best guess at the price. `error` is the difference, and `like` is the likelihood of the data, given the hypothesis.

`ErrorDensity` is defined in `Player`:

```
# class Player:

    def ErrorDensity(self, error):
        return self.pdf_error.Density(error)
```

`ErrorDensity` works by evaluating `pdf_error` at the given value of `error`. The result is a probability density, so it is not really a probability. But remember that `Likelihood` doesn't need to compute a probability; it only has to compute something *proportional* to a probability. As long as the constant of proportionality is the same for all likelihoods, it gets canceled out when we normalize the posterior distribution.

And therefore, a probability density is a perfectly good likelihood.

Update

`Player` provides a method that takes the contestant's guess and computes the posterior distribution:

```
# class Player

    def MakeBeliefs(self, guess):
        pmf = self.PmfPrice()
        self.prior = Price(pmf, self)
        self.posterior = self.prior.Copy()
        self.posterior.Update(guess)
```

`PmfPrice` generates a discrete approximation to the PDF of price, which we use to construct the prior.

`PmfPrice` uses `MakePmf`, which evaluates `pdf_price` at a sequence of values:

```
# class Player

    n = 101
    price_xs = numpy.linspace(0, 75000, n)

    def PmfPrice(self):
        return self.pdf_price.MakePmf(self.price_xs)
```

To construct the posterior, we make a copy of the prior and then invoke `Update`, which invokes `Likelihood` for each hypothesis, multiplies the priors by the likelihoods, and renormalizes.

So let's get back to the original scenario. Suppose you are Player 1 and when you see your showcase, your best guess is that the total price of the prizes is $20,000.

Figure 6-3 shows prior and posterior beliefs about the actual price. The posterior is shifted to the left because your guess is on the low end of the prior range.

On one level, this result makes sense. The most likely value in the prior is $27,750, your best guess is $20,000, and the mean of the posterior is somewhere in between: $25,096.

On another level, you might find this result bizarre, because it suggests that if you *think* the price is $20,000, then you should *believe* the price is $24,000.

To resolve this apparent paradox, remember that you are combining two sources of information, historical data about past showcases and guesses about the prizes you see.

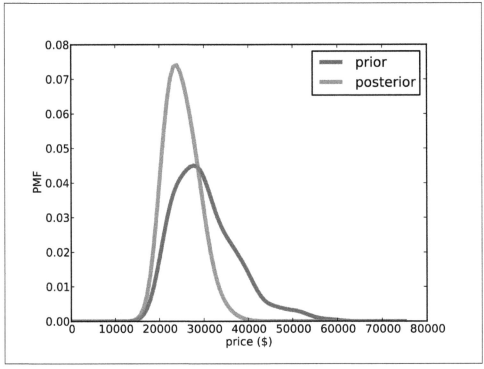

Figure 6-3. Prior and posterior distributions for Player 1, based on a best guess of $20,000.

We are treating the historical data as the prior and updating it based on your guesses, but we could equivalently use your guess as a prior and update it based on historical data.

If you think of it that way, maybe it is less surprising that the most likely value in the posterior is not your original guess.

Optimal bidding

Now that we have a posterior distribution, we can use it to compute the optimal bid, which I define as the bid that maximizes expected return (see *http://en.wikipedia.org/wiki/Expected_return*).

I'm going to present the methods in this section top-down, which means I will show you how they are used before I show you how they work. If you see an unfamiliar method, don't worry; the definition will be along shortly.

To compute optimal bids, I wrote a class called `GainCalculator`:

```
class GainCalculator(object):

    def __init__(self, player, opponent):
        self.player = player
        self.opponent = opponent
```

`player` and `opponent` are `Player` objects.

`GainCalculator` provides `ExpectedGains`, which computes a sequence of bids and the expected gain for each bid:

```
def ExpectedGains(self, low=0, high=75000, n=101):
    bids = numpy.linspace(low, high, n)

    gains = [self.ExpectedGain(bid) for bid in bids]

    return bids, gains
```

`low` and `high` specify the range of possible bids; `n` is the number of bids to try.

`ExpectedGains` calls `ExpectedGain`, which computes expected gain for a given bid:

```
def ExpectedGain(self, bid):
    suite = self.player.posterior
    total = 0
    for price, prob in sorted(suite.Items()):
        gain = self.Gain(bid, price)
        total += prob * gain
    return total
```

`ExpectedGain` loops through the values in the posterior and computes the gain for each bid, given the actual prices of the showcase. It weights each gain with the corresponding probability and returns the total.

`ExpectedGain` invokes `Gain`, which takes a bid and an actual price and returns the expected gain:

```
def Gain(self, bid, price):
    if bid > price:
        return 0

    diff = price - bid
    prob = self.ProbWin(diff)

    if diff <= 250:
        return 2 * price * prob
    else:
        return price * prob
```

If you overbid, you get nothing. Otherwise we compute the difference between your bid and the price, which determines your probability of winning.

If `diff` is less than $250, you win both showcases. For simplicity, I assume that both showcases have the same price. Since this outcome is rare, it doesn't make much difference.

Finally, we have to compute the probability of winning based on `diff`:

```
def ProbWin(self, diff):
    prob = (self.opponent.ProbOverbid() +
            self.opponent.ProbWorseThan(diff))
    return prob
```

If your opponent overbids, you win. Otherwise, you have to hope that your opponent is off by more than `diff`. Player provides methods to compute both probabilities:

```
# class Player:

    def ProbOverbid(self):
        return self.cdf_diff.Prob(-1)

    def ProbWorseThan(self, diff):
        return 1 - self.cdf_diff.Prob(diff)
```

This code might be confusing because the computation is now from the point of view of the opponent, who is computing, "What is the probability that I overbid?" and "What is the probability that my bid is off by more than `diff`?"

Both answers are based on the CDF of `diff`. If the opponent's `diff` is less than or equal to -1, you win. If the opponent's `diff` is worse than yours, you win. Otherwise you lose.

Finally, here's the code that computes optimal bids:

```
# class Player:

    def OptimalBid(self, guess, opponent):
        self.MakeBeliefs(guess)
        calc = GainCalculator(self, opponent)
        bids, gains = calc.ExpectedGains()
        gain, bid = max(zip(gains, bids))
        return bid, gain
```

Given a guess and an opponent, `OptimalBid` computes the posterior distribution, instantiates a `GainCalculator`, computes expected gains for a range of bids and returns the optimal bid and expected gain. Whew!

Figure 6-4 shows the results for both players, based on a scenario where Player 1's best guess is $20,000 and Player 2's best guess is $40,000.

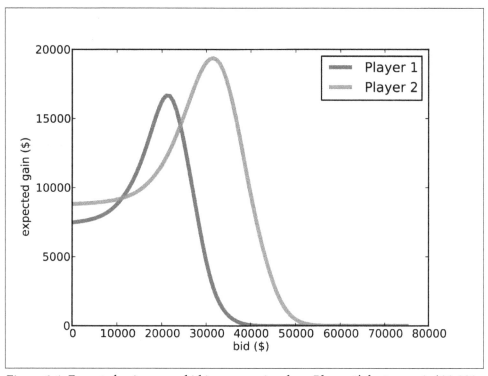

Figure 6-4. Expected gain versus bid in a scenario where Player 1's best guess is $20,000 and Player 2's best guess is $40,000.

For Player 1 the optimal bid is $21,000, yielding an expected return of almost $16,700. This is a case (which turns out to be unusual) where the optimal bid is actually higher than the contestant's best guess.

For Player 2 the optimal bid is $31,500, yielding an expected return of almost $19,400. This is the more typical case where the optimal bid is less than the best guess.

Discussion

One of the features of Bayesian estimation is that the result comes in the form of a posterior distribution. Classical estimation usually generates a single point estimate or a confidence interval, which is sufficient if estimation is the last step in the process, but if you want to use an estimate as an input to a subsequent analysis, point estimates and intervals are often not much help.

In this example, we use the posterior distribution to compute an optimal bid. The return on a given bid is asymmetric and discontinuous (if you overbid, you lose), so it would be hard to solve this problem analytically. But it is relatively simple to do computationally.

Newcomers to Bayesian thinking are often tempted to summarize the posterior distribution by computing the mean or the maximum likelihood estimate. These summaries can be useful, but if that's all you need, then you probably don't need Bayesian methods in the first place.

Bayesian methods are most useful when you can carry the posterior distribution into the next step of the analysis to perform some kind of decision analysis, as we did in this chapter, or some kind of prediction, as we see in the next chapter.

CHAPTER 7
Prediction

The Boston Bruins problem

In the 2010-11 National Hockey League (NHL) Finals, my beloved Boston Bruins played a best-of-seven championship series against the despised Vancouver Canucks. Boston lost the first two games 0-1 and 2-3, then won the next two games 8-1 and 4-0. At this point in the series, what is the probability that Boston will win the next game, and what is their probability of winning the championship?

As always, to answer a question like this, we need to make some assumptions. First, it is reasonable to believe that goal scoring in hockey is at least approximately a Poisson process, which means that it is equally likely for a goal to be scored at any time during a game. Second, we can assume that against a particular opponent, each team has some long-term average goals per game, denoted λ.

Given these assumptions, my strategy for answering this question is

1. Use statistics from previous games to choose a prior distribution for λ.
2. Use the score from the first four games to estimate λ for each team.
3. Use the posterior distributions of λ to compute distribution of goals for each team, the distribution of the goal differential, and the probability that each team wins the next game.
4. Compute the probability that each team wins the series.

To choose a prior distribution, I got some statistics from *http://www.nhl.com*, specifically the average goals per game for each team in the 2010-11 season. The distribution is roughly Gaussian with mean 2.8 and standard deviation 0.3.

The Gaussian distribution is continuous, but we'll approximate it with a discrete Pmf. `thinkbayes` provides `MakeGaussianPmf` to do exactly that:

```
def MakeGaussianPmf(mu, sigma, num_sigmas, n=101):
    pmf = Pmf()
    low = mu - num_sigmas*sigma
    high = mu + num_sigmas*sigma

    for x in numpy.linspace(low, high, n):
        p = scipy.stats.norm.pdf(mu, sigma, x)
        pmf.Set(x, p)
    pmf.Normalize()
    return pmf
```

`mu` and `sigma` are the mean and standard deviation of the Gaussian distribution. `num_sigmas` is the number of standard deviations above and below the mean that the Pmf will span, and `n` is the number of values in the Pmf.

Again we use `numpy.linspace` to make an array of n equally spaced values between `low` and `high`, including both.

`norm.pdf` evaluates the Gaussian probability density function (PDF).

Getting back to the hockey problem, here's the definition for a suite of hypotheses about the value of λ.

```
class Hockey(thinkbayes.Suite):

    def __init__(self):
        pmf = thinkbayes.MakeGaussianPmf(2.7, 0.3, 4)
        thinkbayes.Suite.__init__(self, pmf)
```

So the prior distribution is Gaussian with mean 2.7, standard deviation 0.3, and it spans 4 sigmas above and below the mean.

As always, we have to decide how to represent each hypothesis; in this case I represent the hypothesis that $\lambda = x$ with the floating-point value x.

Poisson processes

In mathematical statistics, a **process** is a stochastic model of a physical system ("stochastic" means that the model has some kind of randomness in it). For example, a Bernoulli process is a model of a sequence of events, called trials, in which each trial has two possible outcomes, like success and failure. So a Bernoulli process is a natural model for a series of coin flips, or a series of shots on goal.

A Poisson process is the continuous version of a Bernoulli process, where an event can occur at any point in time with equal probability. Poisson processes can be used to model customers arriving in a store, buses arriving at a bus stop, or goals scored in a hockey game.

In many real systems the probability of an event changes over time. Customers are more likely to go to a store at certain times of day, buses are supposed to arrive at fixed intervals, and goals are more or less likely at different times during a game.

But all models are based on simplifications, and in this case modeling a hockey game with a Poisson process is a reasonable choice. Heuer, Müller and Rubner (2010) analyze scoring in a German soccer league and come to the same conclusion; see *http://www.cimat.mx/Eventos/vpec10/img/poisson.pdf*.

The benefit of using this model is that we can compute the distribution of goals per game efficiently, as well as the distribution of time between goals. Specifically, if the average number of goals in a game is `lam`, the distribution of goals per game is given by the Poisson PMF:

```
def EvalPoissonPmf(lam, k):
    return (lam)**k * math.exp(-lam) / math.factorial(k)
```

And the distribution of time between goals is given by the exponential PDF:

```
def EvalExponentialPdf(lam, x):
    return lam * math.exp(-lam * x)
```

I use the variable `lam` because `lambda` is a reserved keyword in Python. Both of these functions are in `thinkbayes.py`.

The posteriors

Now we can compute the likelihood that a team with a hypothetical value of `lam` scores k goals in a game:

```
# class Hockey

    def Likelihood(self, data, hypo):
        lam = hypo
        k = data
        like = thinkbayes.EvalPoissonPmf(lam, k)
        return like
```

Each hypothesis is a possible value of λ; `data` is the observed number of goals, k.

With the likelihood function in place, we can make a suite for each team and update them with the scores from the first four games.

```
        suite1 = Hockey('bruins')
        suite1.UpdateSet([0, 2, 8, 4])

        suite2 = Hockey('canucks')
        suite2.UpdateSet([1, 3, 1, 0])
```

Figure 7-1 shows the resulting posterior distributions for lam. Based on the first four games, the most likely values for lam are 2.6 for the Canucks and 2.9 for the Bruins.

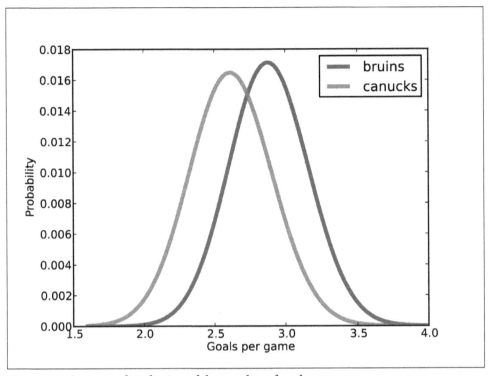

Figure 7-1. Posterior distribution of the number of goals per game.

The distribution of goals

To compute the probability that each team wins the next game, we need to compute the distribution of goals for each team.

If we knew the value of lam exactly, we could use the Poisson distribution again. thinkbayes provides a method that computes a truncated approximation of a Poisson distribution:

```
def MakePoissonPmf(lam, high):
    pmf = Pmf()
    for k in xrange(0, high+1):
        p = EvalPoissonPmf(lam, k)
        pmf.Set(k, p)
    pmf.Normalize()
    return pmf
```

The range of values in the computed Pmf is from 0 to high. So if the value of lam were exactly 3.4, we would compute:

```
lam = 3.4
goal_dist = thinkbayes.MakePoissonPmf(lam, 10)
```

I chose the upper bound, 10, because the probability of scoring more than 10 goals in a game is quite low.

That's simple enough so far; the problem is that we don't know the value of lam exactly. Instead, we have a distribution of possible values for lam.

For each value of lam, the distribution of goals is Poisson. So the overall distribution of goals is a mixture of these Poisson distributions, weighted according to the probabilities in the distribution of lam.

Given the posterior distribution of lam, here's the code that makes the distribution of goals:

```
def MakeGoalPmf(suite):
    metapmf = thinkbayes.Pmf()

    for lam, prob in suite.Items():
        pmf = thinkbayes.MakePoissonPmf(lam, 10)
        metapmf.Set(pmf, prob)

    mix = thinkbayes.MakeMixture(metapmf)
    return mix
```

For each value of lam we make a Poisson Pmf and add it to the meta-Pmf. I call it a meta-Pmf because it is a Pmf that contains Pmfs as its values.

Then we use MakeMixture to compute the mixture (we saw MakeMixture in "Mixtures" on page 50).

Figure 7-2 shows the resulting distribution of goals for the Bruins and Canucks. The Bruins are less likely to score 3 goals or fewer in the next game, and more likely to score 4 or more.

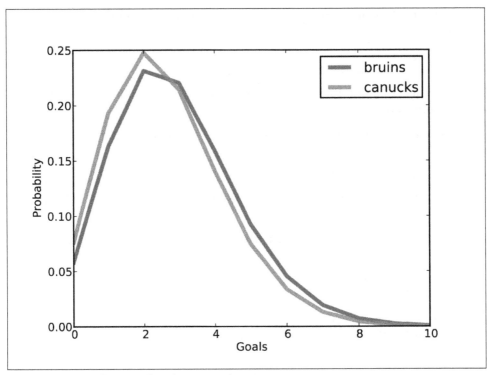

Figure 7-2. Distribution of goals in a single game.

The probability of winning

To get the probability of winning, first we compute the distribution of the goal differential:

```
goal_dist1 = MakeGoalPmf(suite1)
goal_dist2 = MakeGoalPmf(suite2)
diff = goal_dist1 - goal_dist2
```

The subtraction operator invokes `Pmf.__sub__`, which enumerates pairs of values and computes the difference. Subtracting two distributions is almost the same as adding, which we saw in "Addends" on page 44.

If the goal differential is positive, the Bruins win; if negative, the Canucks win; if 0, it's a tie:

```
p_win = diff.ProbGreater(0)
p_loss = diff.ProbLess(0)
p_tie = diff.Prob(0)
```

With the distributions from the previous section, `p_win` is 46%, `p_loss` is 37%, and `p_tie` is 17%.

In the event of a tie at the end of "regulation play," the teams play overtime periods until one team scores. Since the game ends immediately when the first goal is scored, this overtime format is known as "sudden death."

Sudden death

To compute the probability of winning in a sudden death overtime, the important statistic is not goals per game, but time until the first goal. The assumption that goal-scoring is a Poisson process implies that the time between goals is exponentially distributed.

Given lam, we can compute the time between goals like this:

```
lam = 3.4
time_dist = thinkbayes.MakeExponentialPmf(lam, high=2, n=101)
```

high is the upper bound of the distribution. In this case I chose 2, because the probability of going more than two games without scoring is small. n is the number of values in the Pmf.

If we know lam exactly, that's all there is to it. But we don't; instead we have a posterior distribution of possible values. So as we did with the distribution of goals, we make a meta-Pmf and compute a mixture of Pmfs.

```
def MakeGoalTimePmf(suite):
    metapmf = thinkbayes.Pmf()

    for lam, prob in suite.Items():
        pmf = thinkbayes.MakeExponentialPmf(lam, high=2, n=2001)
        metapmf.Set(pmf, prob)

    mix = thinkbayes.MakeMixture(metapmf)
    return mix
```

Figure 7-3 shows the resulting distributions. For time values less than one period (one third of a game), the Bruins are more likely to score. The time until the Canucks score is more likely to be longer.

I set the number of values, n, fairly high in order to minimize the number of ties, since it is not possible for both teams to score simultaneously.

Now we compute the probability that the Bruins score first:

```
time_dist1 = MakeGoalTimePmf(suite1)
time_dist2 = MakeGoalTimePmf(suite2)
p_overtime = thinkbayes.PmfProbLess(time_dist1, time_dist2)
```

For the Bruins, the probability of winning in overtime is 52%.

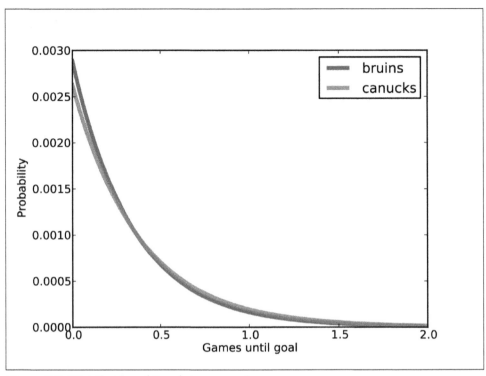

Figure 7-3. Distribution of time between goals.

Finally, the total probability of winning is the chance of winning at the end of regulation play plus the probability of winning in overtime.

```
p_tie = diff.Prob(0)
p_overtime = thinkbayes.PmfProbLess(time_dist1, time_dist2)

p_win = diff.ProbGreater(0) + p_tie * p_overtime
```

For the Bruins, the overall chance of winning the next game is 55%.

To win the series, the Bruins can either win the next two games or split the next two and win the third. Again, we can compute the total probability:

```
# win the next two
p_series = p_win**2

# split the next two, win the third
p_series += 2 * p_win * (1-p_win) * p_win
```

The chance that the Bruins will win the series is 57%. And in 2011, they did.

Discussion

As always, the analysis in this chapter is based on modeling decisions, and modeling is almost always an iterative process. In general, you want to start with something simple that yields an approximate answer, identify likely sources of error, and look for opportunities for improvement.

In this example, I would consider these options:

- I chose a prior based on the average goals per game for each team. But this statistic is averaged across all opponents. Against a particular opponent, we might expect more variability. For example, if the team with the best offense plays the team with the worst defense, the expected goals per game might be several standard deviations above the mean.
- For data I used only the first four games of the championship series. If the same teams played each other during the regular season, I could use the results from those games as well. One complication is that the composition of teams changes during the season due to trades and injuries. So it might be best to give more weight to recent games.
- To take advantage of all available information, we could use results from all regular season games to estimate each team's goal scoring rate, possibly adjusted by estimating an additional factor for each pairwise match-up. This approach would be more complicated, but it is still feasible.

For the first option, we could use the results from the regular season to estimate the variability across all pairwise match-ups. Thanks to Dirk Hoag at *http://forechecker.blogspot.com/*, I was able to get the number of goals scored during regulation play (not overtime) for each game in the regular season.

Teams in different conferences only play each other one or two times in the regular season, so I focused on pairs that played each other 4–6 times. For each pair, I computed the average goals per game, which is an estimate of λ, then plotted the distribution of these estimates.

The mean of these estimates is 2.8, again, but the standard deviation is 0.85, substantially higher than what we got computing one estimate for each team.

If we run the analysis again with the higher-variance prior, the probability that the Bruins win the series is 80%, substantially higher than the result with the low-variance prior, 57%.

So it turns out that the results are sensitive to the prior, which makes sense considering how little data we have to work with. Based on the difference between the low-variance model and the high-variable model, it seems worthwhile to put some effort into getting the prior right.

The code and data for this chapter are available from *http://thinkbayes.com/hockey.py* and *http://thinkbayes.com/hockey_data.csv*. For more information see "Working with the code" on page xi.

Exercises

Exercise 7-1.

If buses arrive at a bus stop every 20 minutes, and you arrive at the bus stop at a random time, your wait time until the bus arrives is uniformly distributed from 0 to 20 minutes.

But in reality, there is variability in the time between buses. Suppose you are waiting for a bus, and you know the historical distribution of time between buses. Compute your distribution of wait times.

Hint: Suppose that the time between buses is either 5 or 10 minutes with equal probability. What is the probability that you arrive during one of the 10 minute intervals?

I solve a version of this problem in the next chapter.

Exercise 7-2.

Suppose that passengers arriving at the bus stop are well-modeled by a Poisson process with parameter λ. If you arrive at the stop and find 3 people waiting, what is your posterior distribution for the time since the last bus arrived.

I solve a version of this problem in the next chapter.

Exercise 7-3.

Suppose that you are an ecologist sampling the insect population in a new environment. You deploy 100 traps in a test area and come back the next day to check on them. You find that 37 traps have been triggered, trapping an insect inside. Once a trap triggers, it cannot trap another insect until it has been reset.

If you reset the traps and come back in two days, how many traps do you expect to find triggered? Compute a posterior predictive distribution for the number of traps.

Exercise 7-4.

Suppose you are the manager of an apartment building with 100 light bulbs in common areas. It is your responsibility to replace light bulbs when they break.

On January 1, all 100 bulbs are working. When you inspect them on February 1, you find 3 light bulbs out. If you come back on April 1, how many light bulbs do you expect to find broken?

In the previous exercise, you could reasonably assume that an event is equally likely at any time. For light bulbs, the likelihood of failure depends on the age of the bulb. Specifically, old bulbs have an increasing failure rate due to evaporation of the filament.

This problem is more open-ended than some; you will have to make modeling decisions. You might want to read about the Weibull distribution (*http://en.wikipedia.org/wiki/Weibull_distribution*). Or you might want to look around for information about light bulb survival curves.

CHAPTER 8
Observer Bias

The Red Line problem

In Massachusetts, the Red Line is a subway that connects Cambridge and Boston. When I was working in Cambridge I took the Red Line from Kendall Square to South Station and caught the commuter rail to Needham. During rush hour Red Line trains run every 7–8 minutes, on average.

When I arrived at the station, I could estimate the time until the next train based on the number of passengers on the platform. If there were only a few people, I inferred that I just missed a train and expected to wait about 7 minutes. If there were more passengers, I expected the train to arrive sooner. But if there were a large number of passengers, I suspected that trains were not running on schedule, so I would go back to the street level and get a taxi.

While I was waiting for trains, I thought about how Bayesian estimation could help predict my wait time and decide when I should give up and take a taxi. This chapter presents the analysis I came up with.

This chapter is based on a project by Brendan Ritter and Kai Austin, who took a class with me at Olin College. The code in this chapter is available from *http://thinkbayes.com/redline.py*. The code I used to collect data is in *http://thinkbayes.com/redline_data.py*. For more information see "Working with the code" on page xi.

The model

Before we get to the analysis, we have to make some modeling decisions. First, I will treat passenger arrivals as a Poisson process, which means I assume that passengers are equally likely to arrive at any time, and that they arrive at an unknown rate, λ, measured in passengers per minute. Since I observe passengers during a short period of time, and at the same time every day, I assume that λ is constant.

On the other hand, the arrival process for trains is not Poisson. Trains to Boston are supposed to leave from the end of the line (Alewife station) every 7–8 minutes during peak times, but by the time they get to Kendall Square, the time between trains varies between 3 and 12 minutes.

To gather data on the time between trains, I wrote a script that downloads real-time data from *http://www.mbta.com/rider_tools/developers/*, selects south-bound trains arriving at Kendall square, and records their arrival times in a database. I ran the script from 4pm to 6pm every weekday for 5 days, and recorded about 15 arrivals per day. Then I computed the time between consecutive arrivals; the distribution of these gaps is shown in Figure 8-1, labeled z.

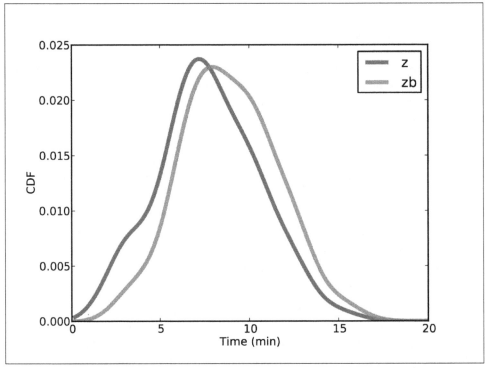

Figure 8-1. PMF of gaps between trains, based on collected data, smoothed by KDE. z is the actual distribution; zb is the biased distribution seen by passengers.

If you stood on the platform from 4pm to 6pm and recorded the time between trains, this is the distribution you would see. But if you arrive at some random time (without regard to the train schedule) you would see a different distribution. The average time between trains, as seen by a random passenger, is substantially higher than the true average.

Why? Because a passenger is more like to arrive during a large interval than a small one. Consider a simple example: suppose that the time between trains is either 5 minutes or 10 minutes with equal probability. In that case the average time between trains is 7.5 minutes.

But a passenger is more likely to arrive during a 10 minute gap than a 5 minute gap; in fact, twice as likely. If we surveyed arriving passengers, we would find that 2/3 of them arrived during a 10 minute gap, and only 1/3 during a 5 minute gap. So the average time between trains, as seen by an arriving passenger, is 8.33 minutes.

This kind of **observer bias** appears in many contexts. Students think that classes are bigger than they are because more of them are in the big classes. Airline passengers think that planes are fuller than they are because more of them are on full flights.

In each case, values from the actual distribution are oversampled in proportion to their value. In the Red Line example, a gap that is twice as big is twice as likely to be observed.

So given the actual distribution of gaps, we can compute the distribution of gaps as seen by passengers. `BiasPmf` does this computation:

```
def BiasPmf(pmf):
    new_pmf = pmf.Copy()

    for x, p in pmf.Items():
        new_pmf.Mult(x, x)

    new_pmf.Normalize()
    return new_pmf
```

`pmf` is the actual distribution; `new_pmf` is the biased distribution. Inside the loop, we multiply the probability of each value, `x`, by the likelihood it will be observed, which is proportional to `x`. Then we normalize the result.

Figure 8-1 shows the actual distribution of gaps, labeled `z`, and the distribution of gaps seen by passengers, labeled `zb` for "z biased".

Wait times

Wait time, which I call `y`, is the time between the arrival of a passenger and the next arrival of a train. Elapsed time, which I call `x`, is the time between the arrival of the previous train and the arrival of a passenger. I chose these definitions so that `zb = x + y`.

Given the distribution of zb, we can compute the distribution of y. I'll start with a simple case and then generalize. Suppose, as in the previous example, that zb is either 5 minutes with probability 1/3, or 10 minutes with probability 2/3.

If we arrive at a random time during a 5 minute gap, y is uniform from 0 to 5 minutes. If we arrive during a 10 minute gap, y is uniform from 0 to 10. So the overall distribution is a mixture of uniform distributions weighted according to the probability of each gap.

The following function takes the distribution of zb and computes the distribution of y:

```
def PmfOfWaitTime(pmf_zb):
    metapmf = thinkbayes.Pmf()
    for gap, prob in pmf_zb.Items():
        uniform = MakeUniformPmf(0, gap)
        metapmf.Set(uniform, prob)

    pmf_y = thinkbayes.MakeMixture(metapmf)
    return pmf_y
```

PmfOfWaitTime makes a meta-Pmf that maps from each uniform distribution to its probability. Then it uses MakeMixture, which we saw in "Mixtures" on page 50, to compute the mixture.

PmfOfWaitTime also uses MakeUniformPmf, defined here:

```
def MakeUniformPmf(low, high):
    pmf = thinkbayes.Pmf()
    for x in MakeRange(low=low, high=high):
        pmf.Set(x, 1)
    pmf.Normalize()
    return pmf
```

low and high are the range of the uniform distribution, (both ends included). Finally, MakeUniformPmf uses MakeRange, defined here:

```
def MakeRange(low, high, skip=10):
    return range(low, high+skip, skip)
```

MakeRange defines a set of possible values for wait time (expressed in seconds). By default it divides the range into 10 second intervals.

To encapsulate the process of computing these distributions, I created a class called WaitTimeCalculator:

```
class WaitTimeCalculator(object):

    def __init__(self, pmf_z):
        self.pmf_z = pmf_z
        self.pmf_zb = BiasPmf(pmf)

        self.pmf_y = self.PmfOfWaitTime(self.pmf_zb)
        self.pmf_x = self.pmf_y
```

The parameter, pmf_z, is the unbiased distribution of z. pmf_zb is the biased distribution of gap time, as seen by passengers.

pmf_y is the distribution of wait time. pmf_x is the distribution of elapsed time, which is the same as the distribution of wait time. To see why, remember that for a particular value of zp, the distribution of y is uniform from 0 to zp. Also

x = zp - y

So the distribution of x is also uniform from 0 to zp.

Figure 8-2 shows the distribution of z, zb, and y based on the data I collected from the Red Line website.

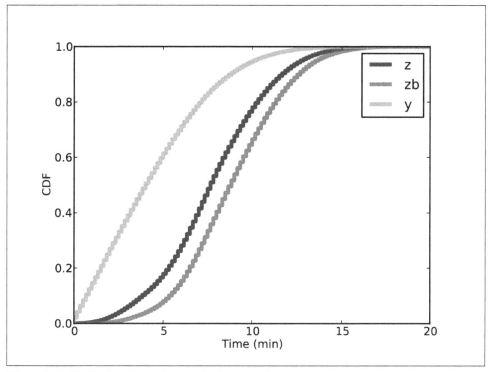

Figure 8-2. CDF of z, zb, and the wait time seen by passengers, y.

To present these distributions, I am switching from Pmfs to Cdfs. Most people are more familiar with Pmfs, but I think Cdfs are easier to interpret, once you get used to them. And if you want to plot several distributions on the same axes, Cdfs are the way to go.

The mean of z is 7.8 minutes. The mean of zb is 8.8 minutes, about 13% higher. The mean of y is 4.4, half the mean of zb.

As an aside, the Red Line schedule reports that trains run every 9 minutes during peak times. This is close to the average of zb, but higher than the average of z. I exchanged email with a representative of the MBTA, who confirmed that the reported time between trains is deliberately conservative in order to account for variability.

Predicting wait times

Let's get back to the motivating question: suppose that when I arrive at the platform I see 10 people waiting. How long should I expect to wait until the next train arrives?

As always, let's start with the easiest version of the problem and work our way up. Suppose we are given the actual distribution of z, and we know that the passenger arrival rate, λ, is 2 passengers per minute.

In that case we can:

1. Use the distribution of z to compute the prior distribution of zp, the time between trains as seen by a passenger.
2. Then we can use the number of passengers to estimate the distribution of x, the elapsed time since the last train.
3. Finally, we use the relation y = zp - x to get the distribution of y.

The first step is to create a `WaitTimeCalculator` that encapsulates the distributions of zp, x, and y, prior to taking into account the number of passengers.

```
wtc = WaitTimeCalculator(pmf_z)
```

pmf_z is the given distribution of gap times.

The next step is to make an `ElapsedTimeEstimator` (defined below), which encapsulates the posterior distribution of x and the predictive distribution of y.

```
ete = ElapsedTimeEstimator(wtc,
                           lam=2.0/60,
                           num_passengers=15)
```

The parameters are the `WaitTimeCalculator`, the passenger arrival rate, `lam` (expressed in passengers per second), and the observed number of passengers, let's say 15.

Here is the definition of `ElapsedTimeEstimator`:

```
class ElapsedTimeEstimator(object):

    def __init__(self, wtc, lam, num_passengers):
        self.prior_x = Elapsed(wtc.pmf_x)

        self.post_x = self.prior_x.Copy()
        self.post_x.Update((lam, num_passengers))

        self.pmf_y = PredictWaitTime(wtc.pmf_zb, self.post_x)
```

`prior_x` and `posterior_x` are the prior and posterior distributions of elapsed time. `pmf_y` is the predictive distribution of wait time.

`ElapsedTimeEstimator` uses `Elapsed` and `PredictWaitTime`, defined below.

`Elapsed` is a Suite that represents the hypothetical distribution of x. The prior distribution of x comes straight from the `WaitTimeCalculator`. Then we use the data, which consists of the arrival rate, `lam`, and the number of passengers on the platform, to compute the posterior distribution.

Here's the definition of `Elapsed`:

```
class Elapsed(thinkbayes.Suite):

    def Likelihood(self, data, hypo):
        x = hypo
        lam, k = data
        like = thinkbayes.EvalPoissonPmf(lam * x, k)
        return like
```

As always, `Likelihood` takes a hypothesis and data, and computes the likelihood of the data under the hypothesis. In this case hypo is the elapsed time since the last train and data is a tuple of `lam` and the number of passengers.

The likelihood of the data is the probability of getting k arrivals in x time, given arrival rate `lam`. We compute that using the PMF of the Poisson distribution.

Finally, here's the definition of `PredictWaitTime`:

```
def PredictWaitTime(pmf_zb, pmf_x):
    pmf_y = pmf_zb - pmf_x
    RemoveNegatives(pmf_y)
    return pmf_y
```

`pmf_zb` is the distribution of gaps between trains; `pmf_x` is the distribution of elapsed time, based on the observed number of passengers. Since y = zb - x, we can compute

```
pmf_y = pmf_zb - pmf_x
```

The subtraction operator invokes `Pmf.__sub__`, which enumerates all pairs of zb and x, computes the differences, and adds the results to pmf_y.

The resulting Pmf includes some negative values, which we know are impossible. For example, if you arrive during a gap of 5 minutes, you can't wait more than 5 minutes. `RemoveNegatives` removes the impossible values from the distribution and renormalizes.

```
def RemoveNegatives(pmf):
    for val in pmf.Values():
        if val < 0:
            pmf.Remove(val)
    pmf.Normalize()
```

Figure 8-3 shows the results. The prior distribution of x is the same as the distribution of y in Figure 8-2. The posterior distribution of x shows that, after seeing 15 passengers on the platform, we believe that the time since the last train is probably 5-10 minutes. The predictive distribution of y indicates that we expect the next train in less than 5 minutes, with about 80% confidence.

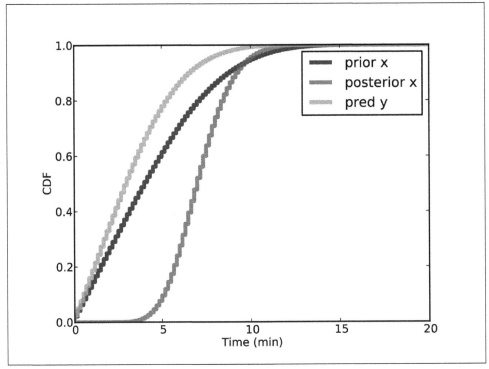

Figure 8-3. Prior and posterior of x and predicted y.

Estimating the arrival rate

The analysis so far has been based on the assumption that we know (1) the distribution of gaps and (2) the passenger arrival rate. Now we are ready to relax the second assumption.

Suppose that you just moved to Boston, so you don't know much about the passenger arrival rate on the Red Line. After a few days of commuting, you could make a guess, at least qualitatively. With a little more effort, you could estimate λ quantitatively.

Each day when you arrive at the platform, you should note the time and the number of passengers waiting (if the platform is too big, you could choose a sample area). Then you should record your wait time and the number of new arrivals while you are waiting.

After five days, you might have data like this:

k1	y	k2
17	4.6	9
22	1.0	0
23	1.4	4
18	5.4	12
4	5.8	11

where k1 is the number of passengers waiting when you arrive, y is your wait time in minutes, and k2 is the number of passengers who arrive while you are waiting.

Over the course of one week, you waited 18 minutes and saw 36 passengers arrive, so you would estimate that the arrival rate is 2 passengers per minute. For practical purposes that estimate is good enough, but for the sake of completeness I will compute a posterior distribution for λ and show how to use that distribution in the rest of the analysis.

ArrivalRate is a Suite that represents hypotheses about λ. As always, Likelihood takes a hypothesis and data, and computes the likelihood of the data under the hypothesis.

In this case the hypothesis is a value of λ. The data is a pair, y, k, where y is a wait time and k is the number of passengers that arrived.

```
class ArrivalRate(thinkbayes.Suite):

    def Likelihood(self, data, hypo):
        lam = hypo
        y, k = data
        like = thinkbayes.EvalPoissonPmf(lam * y, k)
        return like
```

This `Likelihood` might look familiar; it is almost identical to `Elapsed.Likelihood` in "Predicting wait times" on page 84. The difference is that in `Elapsed.Likelihood` the hypothesis is x, the elapsed time; in `ArrivalRate.Likelihood` the hypothesis is `lam`, the arrival rate. But in both cases the likelihood is the probability of seeing k arrivals in some period of time, given `lam`.

`ArrivalRateEstimator` encapsulates the process of estimating λ. The parameter, `passenger_data`, is a list of `k1, y, k2` tuples, as in the table above.

```
class ArrivalRateEstimator(object):

    def __init__(self, passenger_data):
        low, high = 0, 5
        n = 51
        hypos = numpy.linspace(low, high, n) / 60

        self.prior_lam = ArrivalRate(hypos)

        self.post_lam = self.prior_lam.Copy()
        for k1, y, k2 in passenger_data:
            self.post_lam.Update((y, k2))
```

`__init__` builds hypos, which is a sequence of hypothetical values for `lam`, then builds the prior distribution, `prior_lam`. The `for` loop updates the prior with data, yielding the posterior distribution, `post_lam`.

Figure 8-4 shows the prior and posterior distributions. As expected, the mean and median of the posterior are near the observed rate, 2 passengers per minute. But the spread of the posterior distribution captures our uncertainty about λ based on a small sample.

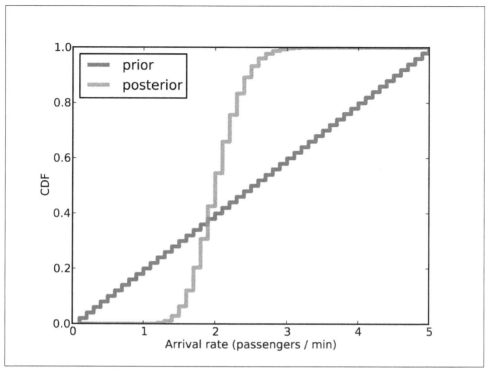

Figure 8-4. Prior and posterior distributions of lam based on five days of passenger data.

Incorporating uncertainty

Whenever there is uncertainty about one of the inputs to an analysis, we can take it into account by a process like this:

1. Implement the analysis based on a deterministic value of the uncertain parameter (in this case λ).
2. Compute the distribution of the uncertain parameter.
3. Run the analysis for each value of the parameter, and generate a set of predictive distributions.
4. Compute a mixture of the predictive distributions, using the weights from the distribution of the parameter.

We have already done steps (1) and (2). I wrote a class called `WaitMixtureEstimator` to handle steps (3) and (4).

```
class WaitMixtureEstimator(object):

    def __init__(self, wtc, are, num_passengers=15):
        self.metapmf = thinkbayes.Pmf()

        for lam, prob in sorted(are.post_lam.Items()):
            ete = ElapsedTimeEstimator(wtc, lam, num_passengers)
            self.metapmf.Set(ete.pmf_y, prob)

        self.mixture = thinkbayes.MakeMixture(self.metapmf)
```

wtc is the WaitTimeCalculator that contains the distribution of zb. are is the ArrivalTimeEstimator that contains the distribution of lam.

The first line makes a meta-Pmf that maps from each possible distribution of y to its probability. For each value of lam, we use ElapsedTimeEstimator to compute the corresponding distribution of y and store it in the Meta-Pmf. Then we use MakeMixture to compute the mixture.

Figure 8-5 shows the results. The shaded lines in the background are the distributions of y for each value of lam, with line thickness that represents likelihood. The dark line is the mixture of these distributions.

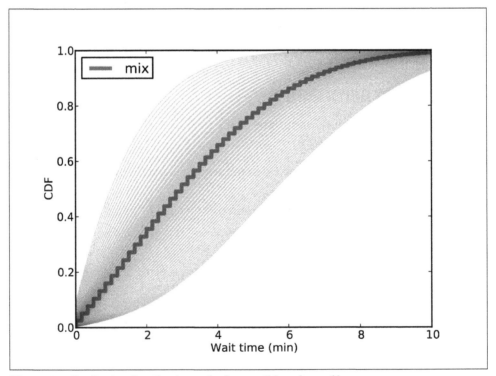

Figure 8-5. Predictive distributions of y for possible values of lam.

In this case we could get a very similar result using a single point estimate of lam. So it was not necessary, for practical purposes, to include the uncertainty of the estimate.

In general, it is important to include variability if the system response is non-linear; that is, if small changes in the input can cause big changes in the output. In this case, posterior variability in lam is small and the system response is approximately linear for small perturbations.

Decision analysis

At this point we can use the number of passengers on the platform to predict the distribution of wait times. Now let's get to the second part of the question: when should I stop waiting for the train and go catch a taxi?

Remember that in the original scenario, I am trying to get to South Station to catch the commuter rail. Suppose I leave the office with enough time that I can wait 15 minutes and still make my connection at South Station.

In that case I would like to know the probability that y exceeds 15 minutes as a function of num_passengers. It is easy enough to use the analysis from "Predicting wait times" on page 84 and run it for a range of num_passengers.

But there's a problem. The analysis is sensitive to the frequency of long delays, and because long delays are rare, it is hard to estimate their frequency.

I only have data from one week, and the longest delay I observed was 15 minutes. So I can't estimate the frequency of longer delays accurately.

However, I can use previous observations to make at least a coarse estimate. When I commuted by Red Line for a year, I saw three long delays caused by a signaling problem, a power outage, and "police activity" at another stop. So I estimate that there are about 3 major delays per year.

But remember that my observations are biased. I am more likely to observe long delays because they affect a large number of passengers. So we should treat my observations as a sample of zb rather than z. Here's how we can do that.

During my year of commuting, I took the Red Line home about 220 times. So I take the observed gap times, gap_times, generate a sample of 220 gaps, and compute their Pmf:

```
n = 220
cdf_z = thinkbayes.MakeCdfFromList(gap_times)
sample_z = cdf_z.Sample(n)
pmf_z = thinkbayes.MakePmfFromList(sample_z)
```

Next I bias pmf_z to get the distribution of zb, draw a sample, and then add in delays of 30, 40, and 50 minutes (expressed in seconds):

```
cdf_zp = BiasPmf(pmf_z).MakeCdf()
sample_zb = cdf_zp.Sample(n) + [1800, 2400, 3000]
```

Cdf.Sample is more efficient than Pmf.Sample, so it is usually faster to convert a Pmf to a Cdf before sampling.

Next I use the sample of zb to estimate a Pdf using KDE, and then convert the Pdf to a Pmf:

```
pdf_zb = thinkbayes.EstimatedPdf(sample_zb)
xs = MakeRange(low=60)
pmf_zb = pdf_zb.MakePmf(xs)
```

Finally I unbias the distribution of zb to get the distribution of z, which I use to create the WaitTimeCalculator:

```
pmf_z = UnbiasPmf(pmf_zb)
wtc = WaitTimeCalculator(pmf_z)
```

This process is complicated, but all of the steps are operations we have seen before. Now we are ready to compute the probability of a long wait.

```
def ProbLongWait(num_passengers, minutes):
    ete = ElapsedTimeEstimator(wtc, lam, num_passengers)
    cdf_y = ete.pmf_y.MakeCdf()
    prob = 1 - cdf_y.Prob(minutes * 60)
```

Given the number of passengers on the platform, ProbLongWait makes an ElapsedTimeEstimator, extracts the distribution of wait time, and computes the probability that wait time exceeds minutes.

Figure 8-6 shows the result. When the number of passengers is less than 20, we infer that the system is operating normally, so the probability of a long delay is small. If there are 30 passengers, we estimate that it has been 15 minutes since the last train; that's longer than a normal delay, so we infer that something is wrong and expect longer delays.

If we are willing to accept a 10% chance of missing the connection at South Station, we should stay and wait as long as there are fewer than 30 passengers, and take a taxi if there are more.

Or, to take this analysis one step further, we could quantify the cost of missing the connection and the cost of taking a taxi, then choose the threshold that minimizes expected cost.

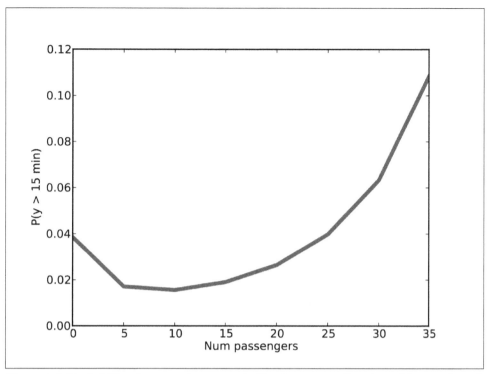

Figure 8-6. Probability that wait time exceeds 15 minutes as a function of the number of passengers on the platform.

Discussion

The analysis so far has been based on the assumption that the arrival rate of passengers is the same every day. For a commuter train during rush hour, that might not be a bad assumption, but there are some obvious exceptions. For example, if there is a special event nearby, a large number of people might arrive at the same time. In that case, the estimate of lam would be too low, so the estimates of x and y would be too high.

If special events are as common as major delays, it would be important to include them in the model. We could do that by extending the distribution of lam to include occasional large values.

We started with the assumption that we know distribution of z. As an alternative, a passenger could estimate z, but it would not be easy. As a passenger, you only observe only your own wait time, y. Unless you skip the first train and wait for the second, you don't observe the gap between trains, z.

However, we could make some inferences about zb. If we note the number of passengers waiting when we arrive, we can estimate the elapsed time since the last train, x. Then we observe y. If we add the posterior distribution of x to the observed y, we get a distribution that represents our posterior belief about the observed value of zb.

We can use this distribution to update our beliefs about the distribution of zb. Finally, we can compute the inverse of `BiasPmf` to get from the distribution of zb to the distribution of z.

I leave this analysis as an exercise for the reader. One suggestion: you should read Chapter 15 first. You can find the outline of a solution in *http://thinkbayes.com/redline.py*. For more information see "Working with the code" on page xi.

Exercises

Exercise 8-1.

This exercise is from MacKay, *Information Theory, Inference, and Learning Algorithms*:

> Unstable particles are emitted from a source and decay at a distance x, a real number that has an exponential probability distribution with [parameter] λ. Decay events can only be observed if they occur in a window extending from $x = 1$ cm to $x = 20$ cm. N decays are observed at locations {1.5, 2, 3, 4, 5, 12} cm. What is the posterior distribution of λ?

You can download a solution to this exercise from *http://thinkbayes.com/decay.py*.

CHAPTER 9
Two Dimensions

Paintball

Paintball is a sport in which competing teams try to shoot each other with guns that fire paint-filled pellets that break on impact, leaving a colorful mark on the target. It is usually played in an arena decorated with barriers and other objects that can be used as cover.

Suppose you are playing paintball in an indoor arena 30 feet wide and 50 feet long. You are standing near one of the 30 foot walls, and you suspect that one of your opponents has taken cover nearby. Along the wall, you see several paint spatters, all the same color, that you think your opponent fired recently.

The spatters are at 15, 16, 18, and 21 feet, measured from the lower-left corner of the room. Based on these data, where do you think your opponent is hiding?

Figure 9-1 shows a diagram of the arena. Using the lower-left corner of the room as the origin, I denote the unknown location of the shooter with coordinates α and β, or `alpha` and `beta`. The location of a spatter is labeled `x`. The angle the opponent shoots at is θ or `theta`.

The Paintball problem is a modified version of the Lighthouse problem, a common example of Bayesian analysis. My notation follows the presentation of the problem in D.S. Sivia's, *Data Analysis: a Bayesian Tutorial, Second Edition* (Oxford, 2006).

You can download the code in this chapter from *http://thinkbayes.com/paintball.py*. For more information see "Working with the code" on page xi.

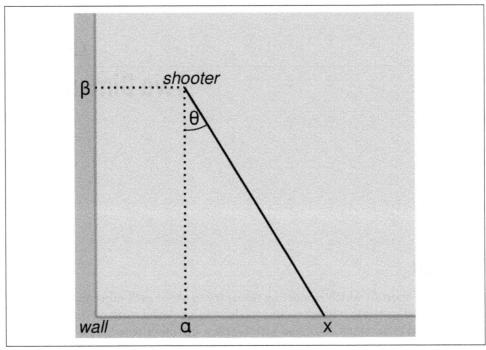

Figure 9-1. Diagram of the layout for the paintball problem.

The suite

To get started, we need a Suite that represents a set of hypotheses about the location of the opponent. Each hypothesis is a pair of coordinates: (`alpha, beta`).

Here is the definition of the Paintball suite:

```
class Paintball(thinkbayes.Suite, thinkbayes.Joint):

    def __init__(self, alphas, betas, locations):
        self.locations = locations
        pairs = [(alpha, beta)
                 for alpha in alphas
                 for beta in betas]
        thinkbayes.Suite.__init__(self, pairs)
```

`Paintball` inherits from `Suite`, which we have seen before, and `Joint`, which I will explain soon.

`alphas` is the list of possible values for `alpha`; `betas` is the list of values for `beta`. `pairs` is a list of all (`alpha, beta`) pairs.

`locations` is a list of possible locations along the wall; it is stored for use in `Likelihood`.

The room is 30 feet wide and 50 feet long, so here's the code that creates the suite:

```
alphas = range(0, 31)
betas = range(1, 51)
locations = range(0, 31)

suite = Paintball(alphas, betas, locations)
```

This prior distribution assumes that all locations in the room are equally likely. Given a map of the room, we might choose a more detailed prior, but we'll start simple.

Trigonometry

Now we need a likelihood function, which means we have to figure out the likelihood of hitting any spot along the wall, given the location of the opponent.

As a simple model, imagine that the opponent is like a rotating turret, equally likely to shoot in any direction. In that case, he is most likely to hit the wall at location `alpha`, and less likely to hit the wall far away from `alpha`.

With a little trigonometry, we can compute the probability of hitting any spot along the wall. Imagine that the shooter fires a shot at angle θ; the pellet would hit the wall at location x, where

$$x - \alpha = \beta \tan \theta$$

Solving this equation for θ yields

$$\theta = \tan^{-1}\left(\frac{x - \alpha}{\beta}\right)$$

So given a location on the wall, we can find θ.

Taking the derivative of the first equation with respect to θ yields

$$\frac{dx}{d\theta} = \frac{\beta}{\cos^2 \theta}$$

This derivative is what I'll call the "strafing speed", which is the speed of the target location along the wall as θ increases. The probability of hitting a given point on the wall is inversely related to strafing speed.

If we know the coordinates of the shooter and a location along the wall, we can compute strafing speed:

```
def StrafingSpeed(alpha, beta, x):
    theta = math.atan2(x - alpha, beta)
    speed = beta / math.cos(theta)**2
    return speed
```

alpha and beta are the coordinates of the shooter; x is the location of a spatter. The result is the derivative of x with respect to theta.

Now we can compute a Pmf that represents the probability of hitting any location on the wall. MakeLocationPmf takes alpha and beta, the coordinates of the shooter, and locations, a list of possible values of x.

```
def MakeLocationPmf(alpha, beta, locations):
    pmf = thinkbayes.Pmf()
    for x in locations:
        prob = 1.0 / StrafingSpeed(alpha, beta, x)
        pmf.Set(x, prob)
    pmf.Normalize()
    return pmf
```

MakeLocationPmf computes the probability of hitting each location, which is inversely related to strafing speed. The result is a Pmf of locations and their probabilities.

Figure 9-2 shows the Pmf of location with alpha = 10 and a range of values for beta. For all values of beta the most likely spatter location is x = 10; as beta increases, so does the spread of the Pmf.

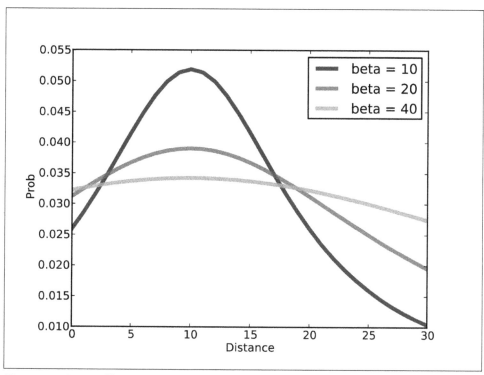

Figure 9-2. PMF of location given alpha=10, for several values of beta.

Likelihood

Now all we need is a likelihood function. We can use `MakeLocationPmf` to compute the likelihood of any value of x, given the coordinates of the opponent.

```
def Likelihood(self, data, hypo):
    alpha, beta = hypo
    x = data
    pmf = MakeLocationPmf(alpha, beta, self.locations)
    like = pmf.Prob(x)
    return like
```

Again, `alpha` and `beta` are the hypothetical coordinates of the shooter, and `x` is the location of an observed spatter.

`pmf` contains the probability of each location, given the coordinates of the shooter. From this Pmf, we select the probability of the observed location.

And we're done. To update the suite, we can use `UpdateSet`, which is inherited from `Suite`.

```
suite.UpdateSet([15, 16, 18, 21])
```

The result is a distribution that maps each (`alpha, beta`) pair to a posterior probability.

Joint distributions

When each value in a distribution is a tuple of variables, it is called a **joint distribution** because it represents the distributions of the variables together, that is "jointly". A joint distribution contains the distributions of the variables, as well information about the relationships among them.

Given a joint distribution, we can compute the distributions of each variable independently, which are called the **marginal distributions**.

`thinkbayes.Joint` provides a method that computes marginal distributions:

```
# class Joint:

    def Marginal(self, i):
        pmf = Pmf()
        for vs, prob in self.Items():
            pmf.Incr(vs[i], prob)
        return pmf
```

`i` is the index of the variable we want; in this example `i=0` indicates the distribution of `alpha`, and `i=1` indicates the distribution of `beta`.

Here's the code that extracts the marginal distributions:

```
marginal_alpha = suite.Marginal(0)
marginal_beta = suite.Marginal(1)
```

Figure 9-3 shows the results (converted to CDFs). The median value for `alpha` is 18, near the center of mass of the observed spatters. For `beta`, the most likely values are close to the wall, but beyond 10 feet the distribution is almost uniform, which indicates that the data do not distinguish strongly between these possible locations.

Given the posterior marginals, we can compute credible intervals for each coordinate independently:

```
print 'alpha CI', marginal_alpha.CredibleInterval(50)
print 'beta CI', marginal_beta.CredibleInterval(50)
```

The 50% credible intervals are (14, 21) for `alpha` and (5, 31) for `beta`. So the data provide evidence that the shooter is in the near side of the room. But it is not strong evidence. The 90% credible intervals cover most of the room!

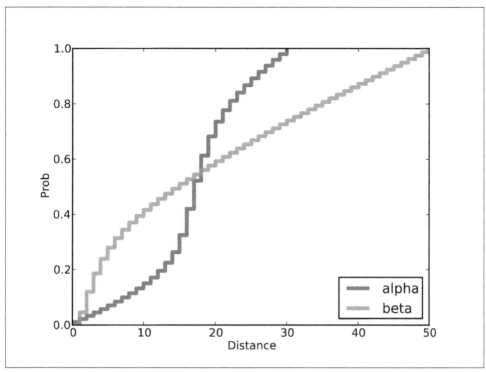

Figure 9-3. Posterior CDFs for alpha and beta, given the data.

Conditional distributions

The marginal distributions contain information about the variables independently, but they do not capture the dependence between variables, if any.

One way to visualize dependence is by computing **conditional distributions**. think bayes.Joint provides a method that does that:

```
def Conditional(self, i, j, val):
    pmf = Pmf()
    for vs, prob in self.Items():
        if vs[j] != val: continue
        pmf.Incr(vs[i], prob)

    pmf.Normalize()
    return pmf
```

Again, `i` is the index of the variable we want; `j` is the index of the conditioning variable, and `val` is the conditional value.

The result is the distribution of the *i*th variable under the condition that the *j*th variable is `val`.

For example, the following code computes the conditional distributions of alpha for a range of values of beta:

```
betas = [10, 20, 40]

for beta in betas:
    cond = suite.Conditional(0, 1, beta)
```

Figure 9-4 shows the results, which we could fully describe as "posterior conditional marginal distributions." Whew!

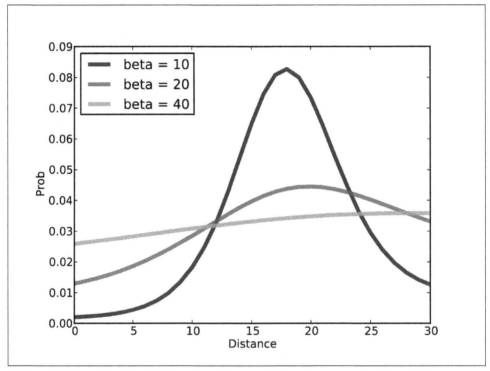

Figure 9-4. Posterior distributions for alpha conditioned on several values of beta.

If the variables were independent, the conditional distributions would all be the same. Since they are all different, we can tell the variables are dependent. For example, if we know (somehow) that beta = 10, the conditional distribution of alpha is fairly narrow. For larger values of beta, the distribution of alpha is wider.

Credible intervals

Another way to visualize the posterior joint distribution is to compute credible intervals. When we looked at credible intervals in "Credible intervals" on page 26, I skipped over a subtle point: for a given distribution, there are many intervals with the same level of credibility. For example, if you want a 50% credible interval, you could choose any set of values whose probability adds up to 50%.

When the values are one-dimensional, it is most common to choose the **central credible interval**; for example, the central 50% credible interval contains all values between the 25th and 75th percentiles.

In multiple dimensions it is less obvious what the right credible interval should be. The best choice might depend on context, but one common choice is the maximum likelihood credible interval, which contains the most likely values that add up to 50% (or some other percentage).

`thinkbayes.Joint` provides a method that computes maximum likelihood credible intervals.

```
# class Joint:

    def MaxLikeInterval(self, percentage=90):
        interval = []
        total = 0

        t = [(prob, val) for val, prob in self.Items()]
        t.sort(reverse=True)

        for prob, val in t:
            interval.append(val)
            total += prob
            if total >= percentage/100.0:
                break

        return interval
```

The first step is to make a list of the values in the suite, sorted in descending order by probability. Next we traverse the list, adding each value to the interval, until the total probability exceeds `percentage`. The result is a list of values from the suite. Notice that this set of values is not necessarily contiguous.

To visualize the intervals, I wrote a function that "colors" each value according to how many intervals it appears in:

```
def MakeCrediblePlot(suite):
    d = dict((pair, 0) for pair in suite.Values())

    percentages = [75, 50, 25]
    for p in percentages:
        interval = suite.MaxLikeInterval(p)
        for pair in interval:
            d[pair] += 1

    return d
```

d is a dictionary that maps from each value in the suite to the number of intervals it appears in. The loop computes intervals for several percentages and modifies d.

Figure 9-5 shows the result. The 25% credible interval is the darkest region near the bottom wall. For higher percentages, the credible interval is bigger, of course, and skewed toward the right side of the room.

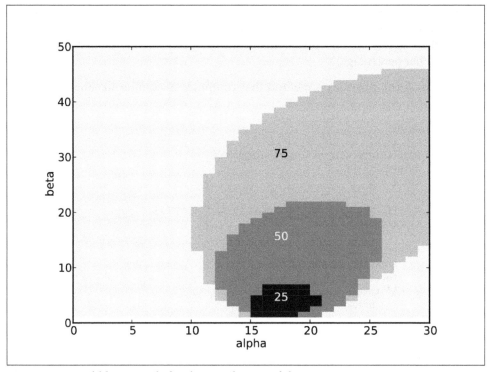

Figure 9-5. Credible intervals for the coordinates of the opponent.

Discussion

This chapter shows that the Bayesian framework from the previous chapters can be extended to handle a two-dimensional parameter space. The only difference is that each hypothesis is represented by a tuple of parameters.

I also presented `Joint`, which is a parent class that provides methods that apply to joint distributions: `Marginal`, `Conditional`, and `MakeLikeInterval`. In object-oriented terms, `Joint` is a mixin (see *http://en.wikipedia.org/wiki/Mixin*).

There is a lot of new vocabulary in this chapter, so let's review:

Joint distribution:
> A distribution that represents all possible values in a multidimensional space and their probabilities. The example in this chapter is a two-dimensional space made up of the coordinates `alpha` and `beta`. The joint distribution represents the probability of each (`alpha`, `beta`) pair.

Marginal distribution:
> The distribution of one parameter in a joint distribution, treating the other parameters as unknown. For example, Figure 9-3 shows the distributions of `alpha` and `beta` independently.

Conditional distribution:
> The distribution of one parameter in a joint distribution, conditioned on one or more of the other parameters. Figure 9-4 several distributions for `alpha`, conditioned on different values of `beta`.

Given the joint distribution, you can compute marginal and conditional distributions. With enough conditional distributions, you could re-create the joint distribution, at least approximately. But given the marginal distributions you cannot re-create the joint distribution because you have lost information about the dependence between variables.

If there are n possible values for each of two parameters, most operations on the joint distribution take time proportional to n^2. If there are d parameters, run time is proportional to n^d, which quickly becomes impractical as the number of dimensions increases.

If you can process a million hypotheses in a reasonable amount of time, you could handle two dimensions with 1000 values for each parameter, or three dimensions with 100 values each, or six dimensions with 10 values each.

If you need more dimensions, or more values per dimension, there are optimizations you can try. I present an example in Chapter 15.

You can download the code in this chapter from *http://thinkbayes.com/paintball.py*. For more information see "Working with the code" on page xi.

Exercises

Exercise 9-1.

In our simple model, the opponent is equally likely to shoot in any direction. As an exercise, let's consider improvements to this model.

The analysis in this chapter suggests that a shooter is most likely to hit the closest wall. But in reality, if the opponent is close to a wall, he is unlikely to shoot at the wall because he is unlikely to see a target between himself and the wall.

Design an improved model that takes this behavior into account. Try to find a model that is more realistic, but not too complicated.

CHAPTER 10
Approximate Bayesian Computation

The Variability Hypothesis

I have a soft spot for crank science. Recently I visited Norumbega Tower, which is an enduring monument to the crackpot theories of Eben Norton Horsford, inventor of double-acting baking powder and fake history. But that's not what this chapter is about.

This chapter is about the Variability Hypothesis, which

> "originated in the early nineteenth century with Johann Meckel, who argued that males have a greater range of ability than females, especially in intelligence. In other words, he believed that most geniuses and most mentally retarded people are men. Because he considered males to be the 'superior animal,' Meckel concluded that females' lack of variation was a sign of inferiority."
>
> From *http://en.wikipedia.org/wiki/Variability_hypothesis*.

I particularly like that last part, because I suspect that if it turns out that women are actually more variable, Meckel would take that as a sign of inferiority, too. Anyway, you will not be surprised to hear that the evidence for the Variability Hypothesis is weak.

Nevertheless, it came up in my class recently when we looked at data from the CDC's Behavioral Risk Factor Surveillance System (BRFSS), specifically the self-reported heights of adult American men and women. The dataset includes responses from 154407 men and 254722 women. Here's what we found:

- The average height for men is 178 cm; the average height for women is 163 cm. So men are taller, on average. No surprise there.
- For men the standard deviation is 7.7 cm; for women it is 7.3 cm. So in absolute terms, men's heights are more variable.

- But to compare variability between groups, it is more meaningful to use the coefficient of variation (CV), which is the standard deviation divided by the mean. It is a dimensionless measure of variability relative to scale. For men CV is 0.0433; for women it is 0.0444.

That's very close, so we could conclude that this dataset provides weak evidence against the Variability Hypothesis. But we can use Bayesian methods to make that conclusion more precise. And answering this question gives me a chance to demonstrate some techniques for working with large datasets.

I will proceed in a few steps:

1. We'll start with the simplest implementation, but it only works for datasets smaller than 1000 values.
2. By computing probabilities under a log transform, we can scale up to the full size of the dataset, but the computation gets slow.
3. Finally, we speed things up substantially with Approximate Bayesian Computation, also known as ABC.

You can download the code in this chapter from *http://thinkbayes.com/variability.py*. For more information see "Working with the code" on page xi.

Mean and standard deviation

In Chapter 9 we estimated two parameters simultaneously using a joint distribution. In this chapter we use the same method to estimate the parameters of a Gaussian distribution: the mean, `mu`, and the standard deviation, `sigma`.

For this problem, I define a Suite called `Height` that represents a map from each `mu`, `sigma` pair to its probability:

```
class Height(thinkbayes.Suite, thinkbayes.Joint):

    def __init__(self, mus, sigmas):

        pairs = [(mu, sigma)
                 for mu in mus
                 for sigma in sigmas]

        thinkbayes.Suite.__init__(self, pairs)
```

`mus` is a sequence of possible values for `mu`; `sigmas` is a sequence of values for `sigma`. The prior distribution is uniform over all `mu`, `sigma` pairs.

The likelihood function is easy. Given hypothetical values of `mu` and `sigma`, we compute the likelihood of a particular value, x. That's what `EvalGaussianPdf` does, so all we have to do is use it:

```
# class Height

    def Likelihood(self, data, hypo):
        x = data
        mu, sigma = hypo
        like = thinkbayes.EvalGaussianPdf(x, mu, sigma)
        return like
```

If you have studied statistics from a mathematical perspective, you know that when you evaluate a PDF, you get a probability density. In order to get a probability, you have to integrate probability densities over some range.

But for our purposes, we don't need a probability; we just need something proportional to the probability we want. A probability density does that job nicely.

The hardest part of this problem turns out to be choosing appropriate ranges for `mus` and `sigmas`. If the range is too small, we omit some possibilities with non-negligible probability and get the wrong answer. If the range is too big, we get the right answer, but waste computational power.

So this is an opportunity to use classical estimation to make Bayesian techniques more efficient. Specifically, we can use classical estimators to find a likely location for `mu` and `sigma`, and use the standard errors of those estimates to choose a likely spread.

If the true parameters of the distribution are μ and σ, and we take a sample of n values, an estimator of μ is the sample mean, m.

And an estimator of σ is the sample standard variance, s.

The standard error of the estimated μ is $s/\sqrt{n}$ and the standard error of the estimated σ is $s/\sqrt{2(n-1)}$.

Here's the code to compute all that:

```
def FindPriorRanges(xs, num_points, num_stderrs=3.0):

    # compute m and s
    n = len(xs)
    m = numpy.mean(xs)
    s = numpy.std(xs)

    # compute ranges for m and s
    stderr_m = s / math.sqrt(n)
    mus = MakeRange(m, stderr_m, num_stderrs)

    stderr_s = s / math.sqrt(2 * (n-1))
    sigmas = MakeRange(s, stderr_s, num_stderrs)

    return mus, sigmas
```

`xs` is the dataset. `num_points` is the desired number of values in the range. `num_stderrs` is the width of the range on each side of the estimate, in number of standard errors.

The return value is a pair of sequences, `mus` and `sigmas`.

Here's `MakeRange`:

```
def MakeRange(estimate, stderr, num_stderrs):
    spread = stderr * num_stderrs
    array = numpy.linspace(estimate-spread,
                           estimate+spread,
                           num_points)
    return array
```

`numpy.linspace` makes an array of equally spaced elements between `estimate-spread` and `estimate+spread`, including both.

Update

Finally here's the code to make and update the suite:

```
mus, sigmas = FindPriorRanges(xs, num_points)
suite = Height(mus, sigmas)
suite.UpdateSet(xs)
print suite.MaximumLikelihood()
```

This process might seem bogus, because we use the data to choose the range of the prior distribution, and then use the data again to do the update. In general, using the same data twice is, in fact, bogus.

But in this case it is ok. Really. We use the data to choose the range for the prior, but only to avoid computing a lot of probabilities that would have been very small anyway. With `num_stderrs=4`, the range is big enough to cover all values with non-negligible likelihood. After that, making it bigger has no effect on the results.

In effect, the prior is uniform over all values of `mu` and `sigma`, but for computational efficiency we ignore all the values that don't matter.

The posterior distribution of CV

Once we have the posterior joint distribution of `mu` and `sigma`, we can compute the distribution of CV for men and women, and then the probability that one exceeds the other.

To compute the distribution of CV, we enumerate pairs of `mu` and `sigma`:

```
def CoefVariation(suite):
    pmf = thinkbayes.Pmf()
    for (mu, sigma), p in suite.Items():
        pmf.Incr(sigma/mu, p)
    return pmf
```

Then we use `thinkbayes.PmfProbGreater` to compute the probability that men are more variable.

The analysis itself is simple, but there are two more issues we have to deal with:

1. As the size of the dataset increases, we run into a series of computational problems due to the limitations of floating-point arithmetic.
2. The dataset contains a number of extreme values that are almost certainly errors. We will need to make the estimation process robust in the presence of these outliers.

The following sections explain these problems and their solutions.

Underflow

If we select the first 100 values from the BRFSS dataset and run the analysis I just described, it runs without errors and we get posterior distributions that look reasonable.

If we select the first 1000 values and run the program again, we get an error in `Pmf.Normalize`:

    ValueError: total probability is zero.

The problem is that we are using probability densities to compute likelihoods, and densities from continuous distributions tend to be small. And if you take 1000 small values and multiply them together, the result is very small. In this case it is so small it can't be represented by a floating-point number, so it gets rounded down to zero, which is called **underflow**. And if all probabilities in the distribution are 0, it's not a distribution any more.

A possible solution is to renormalize the Pmf after each update, or after each batch of 100. That would work, but it would be slow.

A better alternative is to compute likelihoods under a log transform. That way, instead of multiplying small values, we can add up log likelihoods. `Pmf` provides methods `Log`, `LogUpdateSet` and `Exp` to make this process easy.

`Log` computes the log of the probabilities in a Pmf:

```
# class Pmf

    def Log(self):
        m = self.MaxLike()
        for x, p in self.d.iteritems():
            if p:
                self.Set(x, math.log(p/m))
            else:
                self.Remove(x)
```

Before applying the log transform `Log` uses `MaxLike` to find `m`, the highest probability in the Pmf. It divide all probabilities by `m`, so the highest probability gets normalized to 1, which yields a log of 0. The other log probabilities are all negative. If there are any values in the Pmf with probability 0, they are removed.

While the Pmf is under a log transform, we can't use `Update`, `UpdateSet`, or `Normalize`. The result would be nonsensical; if you try, Pmf raises an exception. Instead, we have to use `LogUpdate` and `LogUpdateSet`.

Here's the implementation of `LogUpdateSet`:

```
# class Suite

    def LogUpdateSet(self, dataset):
        for data in dataset:
            self.LogUpdate(data)
```

`LogUpdateSet` loops through the data and calls `LogUpdate`:

```
# class Suite

    def LogUpdate(self, data):
        for hypo in self.Values():
            like = self.LogLikelihood(data, hypo)
            self.Incr(hypo, like)
```

`LogUpdate` is just like `Update` except that it calls `LogLikelihood` instead of `Likelihood`, and `Incr` instead of `Mult`.

Using log-likelihoods avoids the problem with underflow, but while the Pmf is under the log transform, there's not much we can do with it. We have to use `Exp` to invert the transform:

```
# class Pmf

    def Exp(self):
        m = self.MaxLike()
        for x, p in self.d.iteritems():
            self.Set(x, math.exp(p-m))
```

If the log-likelihoods are large negative numbers, the resulting likelihoods might underflow. So `Exp` finds the maximum log-likelihood, `m`, and shifts all the likelihoods up by m. The resulting distribution has a maximum likelihood of 1. This process inverts the log transform with minimal loss of precision.

Log-likelihood

Now all we need is `LogLikelihood`.

```
# class Height

    def LogLikelihood(self, data, hypo):
        x = data
        mu, sigma = hypo
        loglike = scipy.stats.norm.logpdf(x, mu, sigma)
        return loglike
```

`norm.logpdf` computes the log-likelihood of the Gaussian PDF.

Here's what the whole update process looks like:

```
suite.Log()
suite.LogUpdateSet(xs)
suite.Exp()
suite.Normalize()
```

To review, `Log` puts the suite under a log transform. `LogUpdateSet` calls `LogUpdate`, which calls `LogLikelihood`. `LogUpdate` uses `Pmf.Incr`, because adding a log-likelihood is the same as multiplying by a likelihood.

After the update, the log-likelihoods are large negative numbers, so `Exp` shifts them up before inverting the transform, which is how we avoid underflow.

Once the suite is transformed back, the probabilities are "linear" again, which means "not logarithmic", so we can use `Normalize` again.

Using this algorithm, we can process the entire dataset without underflow, but it is still slow. On my computer it might take an hour. We can do better.

A little optimization

This section uses math and computational optimization to speed things up by a factor of 100. But the following section presents an algorithm that is even faster. So if you want to get right to the good stuff, feel free to skip this section.

`Suite.LogUpdateSet` calls `LogUpdate` once for each data point. We can speed it up by computing the log-likelihood of the entire dataset at once.

We'll start with the Gaussian PDF:

$$\frac{1}{\sigma\sqrt{2\pi}} \exp\left[-\frac{1}{2}\left(\frac{x-\mu}{\sigma}\right)^2\right]$$

and compute the log (dropping the constant term):

$$-\log \sigma - \frac{1}{2}\left(\frac{x-\mu}{\sigma}\right)^2$$

Given a sequence of values, x_i, the total log-likelihood is

$$\sum_i -\log \sigma - \frac{1}{2}\left(\frac{x_i-\mu}{\sigma}\right)^2$$

Pulling out the terms that don't depend on *i*, we get

$$-n \log \sigma - \frac{1}{2\sigma^2}\sum_i (x_i-\mu)^2$$

which we can translate into Python:

```
# class Height

    def LogUpdateSetFast(self, data):
        xs = tuple(data)
        n = len(xs)

        for hypo in self.Values():
            mu, sigma = hypo
            total = Summation(xs, mu)
            loglike = -n * math.log(sigma) - total / 2 / sigma**2
            self.Incr(hypo, loglike)
```

By itself, this would be a small improvement, but it creates an opportunity for a bigger one. Notice that the summation only depends on mu, not sigma, so we only have to compute it once for each value of mu.

To avoid recomputing, I factor out a function that computes the summation, and **memoize** it so it stores previously computed results in a dictionary (see http://en.wikipedia.org/wiki/Memoization):

```
def Summation(xs, mu, cache={}):
    try:
        return cache[xs, mu]
    except KeyError:
        ds = [(x-mu)**2 for x in xs]
        total = sum(ds)
        cache[xs, mu] = total
        return total
```

`cache` stores previously computed sums. The `try` statement returns a result from the cache if possible; otherwise it computes the summation, then caches and returns the result.

The only catch is that we can't use a list as a key in the cache, because it is not a hashable type. That's why `LogUpdateSetFast` converts the dataset to a tuple.

This optimization speeds up the computation by about a factor of 100, processing the entire dataset (154407 men and 254722 women) in less than a minute on my not-very-fast computer.

ABC

But maybe you don't have that kind of time. In that case, Approximate Bayesian Computation (ABC) might be the way to go. The motivation behind ABC is that the likelihood of any particular dataset is:

1. Very small, especially for large datasets, which is why we had to use the log transform,
2. Expensive to compute, which is why we had to do so much optimization, and
3. Not really what we want anyway.

We don't really care about the likelihood of seeing the exact dataset we saw. Especially for continuous variables, we care about the likelihood of seeing any dataset like the one we saw.

For example, in the Euro problem, we don't care about the order of the coin flips, only the total number of heads and tails. And in the locomotive problem, we don't care about which particular trains were seen, only the number of trains and the maximum of the serial numbers.

Similarly, in the BRFSS sample, we don't really want to know the probability of seeing one particular set of values (especially since there are hundreds of thousands of them). It is more relevant to ask, "If we sample 100,000 people from a population with hypothetical values of μ and σ, what would be the chance of collecting a sample with the observed mean and variance?"

For samples from a Gaussian distribution, we can answer this question efficiently because we can find the distribution of the sample statistics analytically. In fact, we already did it when we computed the range of the prior.

If you draw n values from a Gaussian distribution with parameters μ and σ, and compute the sample mean, m, the distribution of m is Gaussian with parameters μ and $\sigma/\sqrt{n}$.

Similarly, the distribution of the sample standard deviation, s, is Gaussian with parameters σ and $\sigma/\sqrt{2(n-1)}$.

We can use these sample distributions to compute the likelihood of the sample statistics, m and s, given hypothetical values for μ and σ. Here's a new version of LogUpdateSet that does it:

```python
def LogUpdateSetABC(self, data):
    xs = data
    n = len(xs)

    # compute sample statistics
    m = numpy.mean(xs)
    s = numpy.std(xs)

    for hypo in sorted(self.Values()):
        mu, sigma = hypo

        # compute log likelihood of m, given hypo
        stderr_m = sigma / math.sqrt(n)
        loglike = EvalGaussianLogPdf(m, mu, stderr_m)

        #compute log likelihood of s, given hypo
        stderr_s = sigma / math.sqrt(2 * (n-1))
        loglike += EvalGaussianLogPdf(s, sigma, stderr_s)

        self.Incr(hypo, loglike)
```

On my computer this function processes the entire dataset in about a second, and the result agrees with the exact result with about 5 digits of precision.

Robust estimation

We are almost ready to look at results, but we have one more problem to deal with. There are a number of outliers in this dataset that are almost certainly errors. For example, there are three adults with reported height of 61 cm, which would place them among the shortest living adults in the world. At the other end, there are four women with reported height 229 cm, just short of the tallest women in the world.

It is not impossible that these values are correct, but it is unlikely, which makes it hard to know how to deal with them. And we have to get it right, because these extreme values have a disproportionate effect on the estimated variability.

Because ABC is based on summary statistics, rather than the entire dataset, we can make it more robust by choosing summary statistics that are robust in the presence of outliers. For example, rather than use the sample mean and standard deviation, we could use the median and inter-quartile range (IQR), which is the difference between the 25th and 75th percentiles.

More generally, we could compute an inter-percentile range (IPR) that spans any given fraction of the distribution, p:

```
def MedianIPR(xs, p):
    cdf = thinkbayes.MakeCdfFromList(xs)
    median = cdf.Percentile(50)

    alpha = (1-p) / 2
    ipr = cdf.Value(1-alpha) - cdf.Value(alpha)
    return median, ipr
```

xs is a sequence of values. p is the desired range; for example, p=0.5 yields the interquartile range.

MedianIPR works by computing the CDF of xs, then extracting the median and the difference between two percentiles.

We can convert from ipr to an estimate of sigma using the Gaussian CDF to compute the fraction of the distribution covered by a given number of standard deviations. For example, it is a well-known rule of thumb that 68% of a Gaussian distribution falls within one standard deviation of the mean, which leaves 16% in each tail. If we compute the range between the 16th and 84th percentiles, we expect the result to be 2 * sigma. So we can estimate sigma by computing the 68% IPR and dividing by 2.

More generally we could use any number of sigmas. MedianS performs the more general version of this computation:

```
def MedianS(xs, num_sigmas):
    half_p = thinkbayes.StandardGaussianCdf(num_sigmas) - 0.5

    median, ipr = MedianIPR(xs, half_p * 2)
    s = ipr / 2 / num_sigmas

    return median, s
```

Again, xs is the sequence of values; num_sigmas is the number of standard deviations the results should be based on. The result is median, which estimates μ, and s, which estimates σ.

Finally, in LogUpdateSetABC we can replace the sample mean and standard deviation with median and s. And that pretty much does it.

It might seem odd that we are using observed percentiles to estimate μ and σ, but it is an example of the flexibility of the Bayesian approach. In effect we are asking, "Given hypothetical values for μ and σ, and a sampling process that has some chance of introducing errors, what is the likelihood of generating a given set of sample statistics?"

We are free to choose any sample statistics we like, up to a point: μ and σ determine the location and spread of a distribution, so we need to choose statistics that capture those characteristics. For example, if we chose the 49th and 51st percentiles, we would get very little information about spread, so it would leave the estimate of σ relatively unconstrained by the data. All values of sigma would have nearly the same likelihood of producing the observed values, so the posterior distribution of sigma would look a lot like the prior.

Who is more variable?

Finally we are ready to answer the question we started with: is the coefficient of variation greater for men than for women?

Using ABC based on the median and IPR with num_sigmas=1, I computed posterior joint distributions for mu and sigma. Figures 10-1 and 10-2 show the results as a contour plot with mu on the x-axis, sigma on the y-axis, and probability on the z-axis.

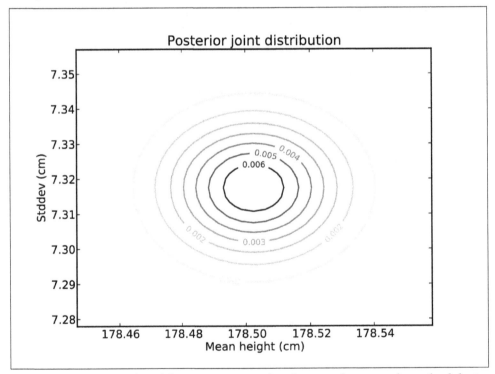

Figure 10-1. Contour plot of the posterior joint distribution of mean and standard deviation of height for men in the U.S.

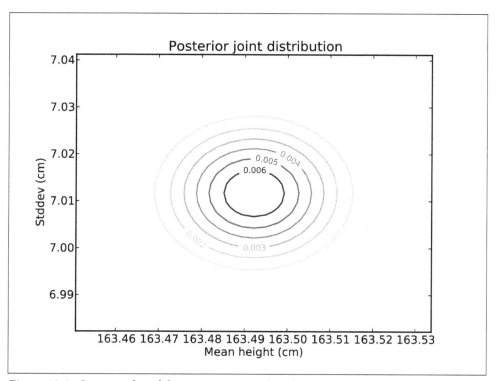

Figure 10-2. Contour plot of the posterior joint distribution of mean and standard deviation of height for women in the U.S.

For each joint distribution, I computed the posterior distribution of CV. Figure 10-3 shows these distributions for men and women. The mean for men is 0.0410; for women it is 0.0429. Since there is no overlap between the distributions, we conclude with near certainty that women are more variable in height than men.

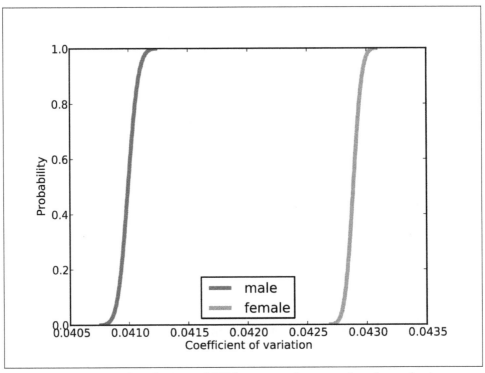

Figure 10-3. Posterior distributions of CV for men and women, based on robust estimators.

So is that the end of the Variability Hypothesis? Sadly, no. It turns out that this result depends on the choice of the inter-percentile range. With num_sigmas=1, we conclude that women are more variable, but with num_sigmas=2 we conclude with equal confidence that men are more variable.

The reason for the difference is that there are more men of short stature, and their distance from the mean is greater.

So our evaluation of the Variability Hypothesis depends on the interpretation of "variability." With num_sigmas=1 we focus on people near the mean. As we increase num_sigmas, we give more weight to the extremes.

To decide which emphasis is appropriate, we would need a more precise statement of the hypothesis. As it is, the Variability Hypothesis may be too vague to evaluate.

Nevertheless, it helped me demonstrate several new ideas and, I hope you agree, it makes an interesting example.

Discussion

There are two ways you might think of ABC. One interpretation is that it is, as the name suggests, an approximation that is faster to compute than the exact value.

But remember that Bayesian analysis is always based on modeling decisions, which implies that there is no "exact" solution. For any interesting physical system there are many possible models, and each model yields different results. To interpret the results, we have to evaluate the models.

So another interpretation of ABC is that it represents an alternative model of the likelihood. When we compute p(D|H), we are asking "What is the likelihood of the data under a given hypothesis?"

For large datasets, the likelihood of the data is very small, which is a hint that we might not be asking the right question. What we really want to know is the likelihood of any outcome like the data, where the definition of "like" is yet another modeling decision.

The underlying idea of ABC is that two datasets are alike if they yield the same summary statistics. But in some cases, like the example in this chapter, it is not obvious which summary statistics to choose.

You can download the code in this chapter from *http://thinkbayes.com/variability.py*. For more information see "Working with the code" on page xi.

Exercises

Exercise 10-1.

An "effect size" is a statistic intended to measure the difference between two groups (see *http://en.wikipedia.org/wiki/Effect_size*).

For example, we could use data from the BRFSS to estimate the difference in height between men and women. By sampling values from the posterior distributions of μ and σ, we could generate the posterior distribution of this difference.

But it might be better to use a dimensionless measure of effect size, rather than a difference measured in cm. One option is to use divide through by the standard deviation (similar to what we did with the coefficient of variation).

If the parameters for Group 1 are (μ_1, σ_1), and the parameters for Group 2 are (μ_2, σ_2), the dimensionless effect size is

$$\frac{\mu_1 - \mu_2}{(\sigma_1 + \sigma_2)/2}$$

Write a function that takes joint distributions of `mu` and `sigma` for two groups and returns the posterior distribution of effect size.

Hint: if enumerating all pairs from the two distributions takes too long, consider random sampling.

CHAPTER 11
Hypothesis Testing

Back to the Euro problem

In "The Euro problem" on page 31 I presented a problem from MacKay's *Information Theory, Inference, and Learning Algorithms*:

> A statistical statement appeared in "The Guardian" on Friday January 4, 2002:
>
>> When spun on edge 250 times, a Belgian one-euro coin came up heads 140 times and tails 110. 'It looks very suspicious to me,' said Barry Blight, a statistics lecturer at the London School of Economics. 'If the coin were unbiased, the chance of getting a result as extreme as that would be less than 7%.'
>
> But do these data give evidence that the coin is biased rather than fair?

We estimated the probability that the coin would land face up, but we didn't really answer MacKay's question: Do the data give evidence that the coin is biased?

In Chapter 4 I proposed that data are in favor of a hypothesis if the data are more likely under the hypothesis than under the alternative or, equivalently, if the Bayes factor is greater than 1.

In the Euro example, we have two hypotheses to consider: I'll use F for the hypothesis that the coin is fair and B for the hypothesis that it is biased.

If the coin is fair, it is easy to compute the likelihood of the data, $p(D|F)$. In fact, we already wrote the function that does it.

```
def Likelihood(self, data, hypo):
    x = hypo / 100.0
    head, tails = data
    like = x**heads * (1-x)**tails
    return like
```

To use it we can create a `Euro` suite and invoke `Likelihood`:

```
suite = Euro()
likelihood = suite.Likelihood(data, 50)
```

$p(D|F)$ is $5.5 \cdot 10^{-76}$, which doesn't tell us much except that the probability of seeing any particular dataset is very small. It takes two likelihoods to make a ratio, so we also have to compute $p(D|B)$.

It is not obvious how to compute the likelihood of B, because it's not obvious what "biased" means.

One possibility is to cheat and look at the data before we define the hypothesis. In that case we would say that "biased" means that the probability of heads is 140/250.

```
actual_percent = 100.0 * 140 / 250
likelihood = suite.Likelihood(data, actual_percent)
```

This version of B I call B_cheat; the likelihood of b_cheat is $34 \cdot 10^{-76}$ and the likelihood ratio is 6.1. So we would say that the data are evidence in favor of this version of B.

But using the data to formulate the hypothesis is obviously bogus. By that definition, any dataset would be evidence in favor of B, unless the observed percentage of heads is exactly 50%.

Making a fair comparison

To make a legitimate comparison, we have to define B without looking at the data. So let's try a different definition. If you inspect a Belgian Euro coin, you might notice that the "heads" side is more prominent than the "tails" side. You might expect the shape to have some effect on x, but be unsure whether it makes heads more or less likely. So you might say "I think the coin is biased so that x is either 0.6 or 0.4, but I am not sure which."

We can think of this version, which I'll call B_two as a hypothesis made up of two sub-hypotheses. We can compute the likelihood for each sub-hypothesis and then compute the average likelihood.

```
like40 = suite.Likelihood(data, 40)
like60 = suite.Likelihood(data, 60)
likelihood = 0.5 * like40 + 0.5 * like60
```

The likelihood ratio (or Bayes factor) for b_two is 1.3, which means the data provide weak evidence in favor of b_two.

More generally, suppose you suspect that the coin is biased, but you have no clue about the value of x. In that case you might build a Suite, which I call b_uniform, to represent sub-hypotheses from 0 to 100.

```
b_uniform = Euro(xrange(0, 101))
b_uniform.Remove(50)
b_uniform.Normalize()
```

I initialize b_uniform with values from 0 to 100. I removed the sub-hypothesis that x is 50%, because if x is 50% the coin is fair, but it has almost no effect on the result whether you remove it or not.

To compute the likelihood of b_uniform we compute the likelihood of each sub-hypothesis and accumulate a weighted average.

```
def SuiteLikelihood(suite, data):
    total = 0
    for hypo, prob in suite.Items():
        like = suite.Likelihood(data, hypo)
        total += prob * like
    return total
```

The likelihood ratio for b_uniform is 0.47, which means that the data are weak evidence against b_uniform, compared to *F*.

If you think about the computation performed by SuiteLikelihood, you might notice that it is similar to an update. To refresh your memory, here's the Update function:

```
def Update(self, data):
    for hypo in self.Values():
        like = self.Likelihood(data, hypo)
        self.Mult(hypo, like)
    return self.Normalize()
```

And here's Normalize:

```
def Normalize(self):
    total = self.Total()

    factor = 1.0 / total
    for x in self.d:
        self.d[x] *= factor

    return total
```

The return value from Normalize is the total of the probabilities in the Suite, which is the average of the likelihoods for the sub-hypotheses, weighted by the prior probabilities. And Update passes this value along, so instead of using SuiteLikelihood, we could compute the likelihood of b_uniform like this:

```
likelihood = b_uniform.Update(data)
```

The triangle prior

In Chapter 4 we also considered a triangle-shaped prior that gives higher probability to values of x near 50%. If we think of this prior as a suite of sub-hypotheses, we can compute its likelihood like this:

```
b_triangle = TrianglePrior()
likelihood = b_triangle.Update(data)
```

The likelihood ratio for `b_triangle` is 0.84, compared to F, so again we would say that the data are weak evidence against B.

The following table shows the priors we have considered, the likelihood of each, and the likelihood ratio (or Bayes factor) relative to F.

Hypothesis	Likelihood $\times 10^{-76}$	Bayes Factor
F	5.5	–
B_cheat	34	6.1
B_two	7.4	1.3
B_uniform	2.6	0.47
B_triangle	4.6	0.84

Depending on which definition we choose, the data might provide evidence for or against the hypothesis that the coin is biased, but in either case it is relatively weak evidence.

In summary, we can use Bayesian hypothesis testing to compare the likelihood of F and B, but we have to do some work to specify precisely what B means. This specification depends on background information about coins and their behavior when spun, so people could reasonably disagree about the right definition.

My presentation of this example follows David MacKay's discussion, and comes to the same conclusion. You can download the code I used in this chapter from *http://thinkbayes.com/euro3.py*. For more information see "Working with the code" on page xi.

Discussion

The Bayes factor for `B_uniform` is 0.47, which means that the data provide evidence against this hypothesis, compared to F. In the previous section I characterized this evidence as "weak," but didn't say why.

Part of the answer is historical. Harold Jeffreys, an early proponent of Bayesian statistics, suggested a scale for interpreting Bayes factors:

Bayes Factor	Strength
1 – 3	Barely worth mentioning
3 – 10	Substantial
10 – 30	Strong
30 – 100	Very strong
> 100	Decisive

In the example, the Bayes factor is 0.47 in favor of `B_uniform`, so it is 2.1 in favor is *F*, which Jeffreys would consider "barely worth mentioning." Other authors have suggested variations on the wording. To avoid arguing about adjectives, we could think about odds instead.

If your prior odds are 1:1, and you see evidence with Bayes factor 2, your posterior odds are 2:1. In terms of probability, the data changed your degree of belief from 50% to 66%. For most real world problems, that change would be small relative to modeling errors and other sources of uncertainty.

On the other hand, if you had seen evidence with Bayes factor 100, your posterior odds would be 100:1 or more than 99%. Whether or not you agree that such evidence is "decisive," it is certainly strong.

Exercises

Exercise 11-1.

Some people believe in the existence of extra-sensory perception (ESP); for example, the ability of some people to guess the value of an unseen playing card with probability better than chance.

What is your prior degree of belief in this kind of ESP? Do you think it is as likely to exist as not? Or are you more skeptical about it? Write down your prior odds.

Now compute the strength of the evidence it would take to convince you that ESP is at least 50% likely to exist. What Bayes factor would be needed to make you 90% sure that ESP exists?

Exercise 11-2.

Suppose that your answer to the previous question is 1000; that is, evidence with Bayes factor 1000 in favor of ESP would be sufficient to change your mind.

Now suppose that you read a paper in a respectable peer-reviewed scientific journal that presents evidence with Bayes factor 1000 in favor of ESP. Would that change your mind?

If not, how do you resolve the apparent contradiction? You might find it helpful to read about David Hume's article, "Of Miracles," at *http://en.wikipedia.org/wiki/Of_Miracles*.

CHAPTER 12
Evidence

Interpreting SAT scores

Suppose you are the Dean of Admission at a small engineering college in Massachusetts, and you are considering two candidates, Alice and Bob, whose qualifications are similar in many ways, with the exception that Alice got a higher score on the Math portion of the SAT, a standardized test intended to measure preparation for college-level work in mathematics.

If Alice got 780 and Bob got a 740 (out of a possible 800), you might want to know whether that difference is evidence that Alice is better prepared than Bob, and what the strength of that evidence is.

Now in reality, both scores are very good, and both candidates are probably well prepared for college math. So the real Dean of Admission would probably suggest that we choose the candidate who best demonstrates the other skills and attitudes we look for in students. But as an example of Bayesian hypothesis testing, let's stick with a narrower question: "How strong is the evidence that Alice is better prepared than Bob?"

To answer that question, we need to make some modeling decisions. I'll start with a simplification I know is wrong; then we'll come back and improve the model. I pretend, temporarily, that all SAT questions are equally difficult. Actually, the designers of the SAT choose questions with a range of difficulty, because that improves the ability to measure statistical differences between test-takers.

But if we choose a model where all questions are equally difficult, we can define a characteristic, p_correct, for each test-taker, which is the probability of answering any question correctly. This simplification makes it easy to compute the likelihood of a given score.

The scale

In order to understand SAT scores, we have to understand the scoring and scaling process. Each test-taker gets a raw score based on the number of correct and incorrect questions. The raw score is converted to a scaled score in the range 200–800.

In 2009, there were 54 questions on the math SAT. The raw score for each test-taker is the number of questions answered correctly minus a penalty of 1/4 point for each question answered incorrectly.

The College Board, which administers the SAT, publishes the map from raw scores to scaled scores. I have downloaded that data and wrapped it in an Interpolator object that provides a forward lookup (from raw score to scaled) and a reverse lookup (from scaled score to raw).

You can download the code for this example from *http://thinkbayes.com/sat.py*. For more information see "Working with the code" on page xi.

The prior

The College Board also publishes the distribution of scaled scores for all test-takers. If we convert each scaled score to a raw score, and divide by the number of questions, the result is an estimate of p_correct. So we can use the distribution of raw scores to model the prior distribution of p_correct.

Here is the code that reads and processes the data:

```
class Exam(object):

    def __init__(self):
        self.scale = ReadScale()
        scores = ReadRanks()
        score_pmf = thinkbayes.MakePmfFromDict(dict(scores))
        self.raw = self.ReverseScale(score_pmf)
        self.prior = DivideValues(raw, 54)
```

Exam encapsulates the information we have about the exam. ReadScale and ReadRanks read files and return objects that contain the data: self.scale is the Interpolator that converts from raw to scaled scores and back; scores is a list of (score, frequency) pairs.

score_pmf is the Pmf of scaled scores. self.raw is the Pmf of raw scores, and self.prior is the Pmf of p_correct.

Figure 12-1 shows the prior distribution of p_correct. This distribution is approximately Gaussian, but it is compressed at the extremes. By design, the SAT has the most power to discriminate between test-takers within two standard deviations of the mean, and less power outside that range.

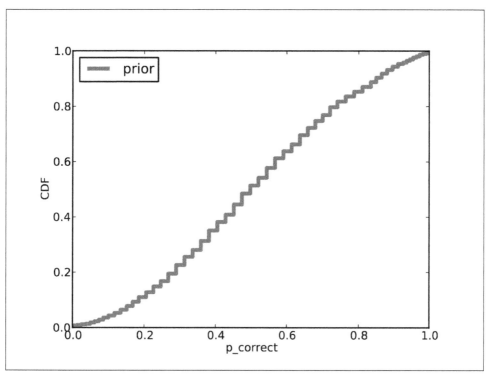

Figure 12-1. Prior distribution of p_correct for SAT test-takers.

For each test-taker, I define a Suite called Sat that represents the distribution of p_correct. Here's the definition:

```
class Sat(thinkbayes.Suite):

    def __init__(self, exam, score):
        thinkbayes.Suite.__init__(self)

        self.exam = exam
        self.score = score

        # start with the prior distribution
        for p_correct, prob in exam.prior.Items():
            self.Set(p_correct, prob)

        # update based on an exam score
        self.Update(score)
```

__init__ takes an Exam object and a scaled score. It makes a copy of the prior distribution and then updates itself based on the exam score.

As usual, we inherit Update from Suite and provide Likelihood:

```
def Likelihood(self, data, hypo):
    p_correct = hypo
    score = data

    k = self.exam.Reverse(score)
    n = self.exam.max_score
    like = thinkbayes.EvalBinomialPmf(k, n, p_correct)
    return like
```

hypo is a hypothetical value of p_correct, and data is a scaled score.

To keep things simple, I interpret the raw score as the number of correct answers, ignoring the penalty for wrong answers. With this simplification, the likelihood is given by the binomial distribution, which computes the probability of k correct responses out of n questions.

Posterior

Figure 12-2 shows the posterior distributions of p_correct for Alice and Bob based on their exam scores. We can see that they overlap, so it is possible that p_correct is actually higher for Bob, but it seems unlikely.

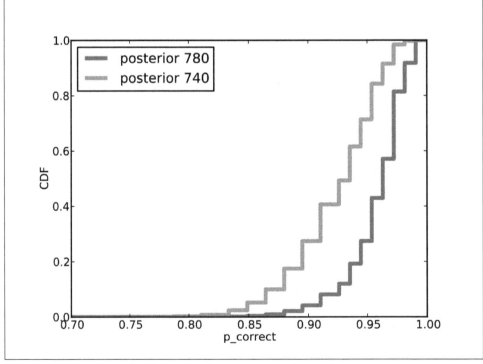

Figure 12-2. Posterior distributions of p_correct for Alice and Bob.

Which brings us back to the original question, "How strong is the evidence that Alice is better prepared than Bob?" We can use the posterior distributions of p_correct to answer this question.

To formulate the question in terms of Bayesian hypothesis testing, I define two hypotheses:

- *A*: p_correct is higher for Alice than for Bob.
- *B*: p_correct is higher for Bob than for Alice.

To compute the likelihood of *A*, we can enumerate all pairs of values from the posterior distributions and add up the total probability of the cases where p_correct is higher for Alice than for Bob. And we already have a function, thinkbayes.PmfProbGreater, that does that.

So we can define a Suite that computes the posterior probabilities of *A* and *B*:

```
class TopLevel(thinkbayes.Suite):

    def Update(self, data):
        a_sat, b_sat = data

        a_like = thinkbayes.PmfProbGreater(a_sat, b_sat)
        b_like = thinkbayes.PmfProbLess(a_sat, b_sat)
        c_like = thinkbayes.PmfProbEqual(a_sat, b_sat)

        a_like += c_like / 2
        b_like += c_like / 2

        self.Mult('A', a_like)
        self.Mult('B', b_like)

        self.Normalize()
```

Usually when we define a new Suite, we inherit Update and provide Likelihood. In this case I override Update, because it is easier to evaluate the likelihood of both hypotheses at the same time.

The data passed to Update are Sat objects that represent the posterior distributions of p_correct.

a_like is the total probability that p_correct is higher for Alice; b_like is that probability that it is higher for Bob.

c_like is the probability that they are "equal," but this equality is an artifact of the decision to model p_correct with a set of discrete values. If we use more values, c_like is smaller, and in the extreme, if p_correct is continuous, c_like is zero. So I treat `c_like` as a kind of round-off error and split it evenly between `a_like` and `b_like`.

Here is the code that creates `TopLevel` and updates it:

```
exam = Exam()
a_sat = Sat(exam, 780)
b_sat = Sat(exam, 740)

top = TopLevel('AB')
top.Update((a_sat, b_sat))
top.Print()
```

The likelihood of *A* is 0.79 and the likelihood of *B* is 0.21. The likelihood ratio (or Bayes factor) is 3.8, which means that these test scores are evidence that Alice is better than Bob at answering SAT questions. If we believed, before seeing the test scores, that *A* and *B* were equally likely, then after seeing the scores we should believe that the probability of *A* is 79%, which means there is still a 21% chance that Bob is actually better prepared.

A better model

Remember that the analysis we have done so far is based on the simplification that all SAT questions are equally difficult. In reality, some are easier than others, which means that the difference between Alice and Bob might be even smaller.

But how big is the modeling error? If it is small, we conclude that the first model—based on the simplification that all questions are equally difficult—is good enough. If it's large, we need a better model.

In the next few sections, I develop a better model and discover (spoiler alert!) that the modeling error is small. So if you are satisfied with the simple model, you can skip to the next chapter. If you want to see how the more realistic model works, read on...

- Assume that each test-taker has some degree of `efficacy`, which measures their ability to answer SAT questions.
- Assume that each question has some level of `difficulty`.
- Finally, assume that the chance that a test-taker answers a question correctly is related to `efficacy` and `difficulty` according to this function:

```
def ProbCorrect(efficacy, difficulty, a=1):
    return 1 / (1 + math.exp(-a * (efficacy - difficulty)))
```

This function is a simplified version of the curve used in **item response theory**, which you can read about at *http://en.wikipedia.org/wiki/Item_response_theory*. efficacy and difficulty are considered to be on the same scale, and the probability of getting a question right depends only on the difference between them.

When efficacy and difficulty are equal, the probability of getting the question right is 50%. As efficacy increases, this probability approaches 100%. As it decreases (or as difficulty increases), the probability approaches 0%.

Given the distribution of efficacy across test-takers and the distribution of difficulty across questions, we can compute the expected distribution of raw scores. We'll do that in two steps. First, for a person with given efficacy, we'll compute the distribution of raw scores.

```
def PmfCorrect(efficacy, difficulties):
    pmf0 = thinkbayes.Pmf([0])

    ps = [ProbCorrect(efficacy, diff) for diff in difficulties]
    pmfs = [BinaryPmf(p) for p in ps]
    dist = sum(pmfs, pmf0)
    return dist
```

difficulties is a list of difficulties, one for each question. ps is a list of probabilities, and pmfs is a list of two-valued Pmf objects; here's the function that makes them:

```
def BinaryPmf(p):
    pmf = thinkbayes.Pmf()
    pmf.Set(1, p)
    pmf.Set(0, 1-p)
    return pmf
```

dist is the sum of these Pmfs. Remember from "Addends" on page 44 that when we add up Pmf objects, the result is the distribution of the sums. In order to use Python's sum to add up Pmfs, we have to provide pmf0 which is the identity for Pmfs, so pmf + pmf0 is always pmf.

If we know a person's efficacy, we can compute their distribution of raw scores. For a group of people with a different efficacies, the resulting distribution of raw scores is a mixture. Here's the code that computes the mixture:

```
# class Exam:

    def MakeRawScoreDist(self, efficacies):
        pmfs = thinkbayes.Pmf()
        for efficacy, prob in efficacies.Items():
            scores = PmfCorrect(efficacy, self.difficulties)
            pmfs.Set(scores, prob)

        mix = thinkbayes.MakeMixture(pmfs)
        return mix
```

`MakeRawScoreDist` takes `efficacies`, which is a Pmf that represents the distribution of efficacy across test-takers. I assume it is Gaussian with mean 0 and standard deviation 1.5. This choice is mostly arbitrary. The probability of getting a question correct depends on the difference between efficacy and difficulty, so we can choose the units of efficacy and then calibrate the units of difficulty accordingly.

`pmfs` is a meta-Pmf that contains one Pmf for each level of efficacy, and maps to the fraction of test-takers at that level. `MakeMixture` takes the meta-pmf and computes the distribution of the mixture (see "Mixtures" on page 50).

Calibration

If we were given the distribution of difficulty, we could use `MakeRawScoreDist` to compute the distribution of raw scores. But for us the problem is the other way around: we are given the distribution of raw scores and we want to infer the distribution of difficulty.

I assume that the distribution of difficulty is uniform with parameters `center` and `width`. `MakeDifficulties` makes a list of difficulties with these parameters.

```
def MakeDifficulties(center, width, n):
    low, high = center-width, center+width
    return numpy.linspace(low, high, n)
```

By trying out a few combinations, I found that `center=-0.05` and `width=1.8` yield a distribution of raw scores similar to the actual data, as shown in Figure 12-3.

So, assuming that the distribution of difficulty is uniform, its range is approximately `-1.85` to `1.75`, given that efficacy is Gaussian with mean 0 and standard deviation 1.5.

The following table shows the range of `ProbCorrect` for test-takers at different levels of efficacy:

Efficacy	Difficulty		
	-1.85	-0.05	1.75
3.00	0.99	0.95	0.78
1.50	0.97	0.82	0.44
0.00	0.86	0.51	0.15
-1.50	0.59	0.19	0.04
-3.00	0.24	0.05	0.01

Someone with efficacy 3 (two standard deviations above the mean) has a 99% chance of answering the easiest questions on the exam, and a 78% chance of answering the hardest. On the other end of the range, someone two standard deviations below the mean has only a 24% chance of answering the easiest questions.

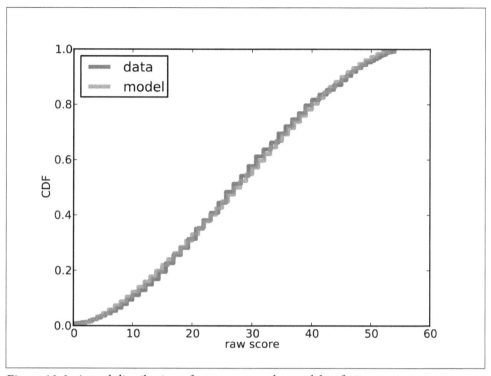

Figure 12-3. Actual distribution of raw scores and a model to fit it.

Posterior distribution of efficacy

Now that the model is calibrated, we can compute the posterior distribution of efficacy for Alice and Bob. Here is a version of the Sat class that uses the new model:

```
class Sat2(thinkbayes.Suite):

    def __init__(self, exam, score):
        self.exam = exam
        self.score = score

        # start with the Gaussian prior
        efficacies = thinkbayes.MakeGaussianPmf(0, 1.5, 3)
        thinkbayes.Suite.__init__(self, efficacies)

        # update based on an exam score
        self.Update(score)
```

`Update` invokes `Likelihood`, which computes the likelihood of a given test score for a hypothetical level of efficacy.

```
def Likelihood(self, data, hypo):
    efficacy = hypo
    score = data
    raw = self.exam.Reverse(score)

    pmf = self.exam.PmfCorrect(efficacy)
    like = pmf.Prob(raw)
    return like
```

`pmf` is the distribution of raw scores for a test-taker with the given efficacy; `like` is the probability of the observed score.

Figure 12-4 shows the posterior distributions of efficacy for Alice and Bob. As expected, the location of Alice's distribution is farther to the right, but again there is some overlap.

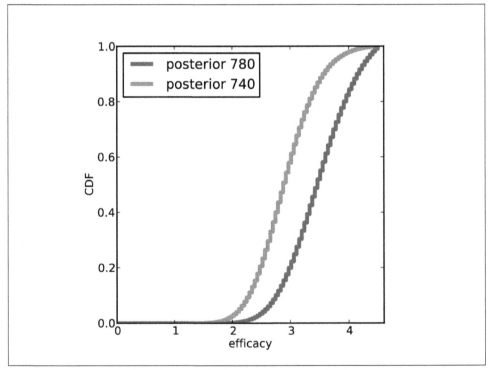

Figure 12-4. Posterior distributions of efficacy for Alice and Bob.

Using `TopLevel` again, we compare A, the hypothesis that Alice's efficacy is higher, and B, the hypothesis that Bob's is higher. The likelihood ratio is 3.4, a bit smaller than what we got from the simple model (3.8). So this model indicates that the data are evidence in favor of *A*, but a little weaker than the previous estimate.

If our prior belief is that *A* and *B* are equally likely, then in light of this evidence we would give *A* a posterior probability of 77%, leaving a 23% chance that Bob's efficacy is higher.

Predictive distribution

The analysis we have done so far generates estimates for Alice and Bob's efficacy, but since efficacy is not directly observable, it is hard to validate the results.

To give the model predictive power, we can use it to answer a related question: "If Alice and Bob take the math SAT again, what is the chance that Alice will do better again?"

We'll answer this question in two steps:

- We'll use the posterior distribution of efficacy to generate a predictive distribution of raw score for each test-taker.
- We'll compare the two predictive distributions to compute the probability that Alice gets a higher score again.

We already have most of the code we need. To compute the predictive distributions, we can use `MakeRawScoreDist` again:

```
exam = Exam()
a_sat = Sat(exam, 780)
b_sat = Sat(exam, 740)

a_pred = exam.MakeRawScoreDist(a_sat)
b_pred = exam.MakeRawScoreDist(b_sat)
```

Then we can find the likelihood that Alice does better on the second test, Bob does better, or they tie:

```
a_like = thinkbayes.PmfProbGreater(a_pred, b_pred)
b_like = thinkbayes.PmfProbLess(a_pred, b_pred)
c_like = thinkbayes.PmfProbEqual(a_pred, b_pred)
```

The probability that Alice does better on the second exam is 63%, which means that Bob has a 37% chance of doing as well or better.

Notice that we have more confidence about Alice's efficacy than we do about the outcome of the next test. The posterior odds are 3:1 that Alice's efficacy is higher, but only 2:1 that Alice will do better on the next exam.

Discussion

We started this chapter with the question, "How strong is the evidence that Alice is better prepared than Bob?" On the face of it, that sounds like we want to test two hypotheses: either Alice is more prepared or Bob is.

But in order to compute likelihoods for these hypotheses, we have to solve an estimation problem. For each test-taker we have to find the posterior distribution of either p_correct or efficacy.

Values like this are called **nuisance parameters** because we don't care what they are, but we have to estimate them to answer the question we care about.

One way to visualize the analysis we did in this chapter is to plot the space of these parameters. thinkbayes.MakeJoint takes two Pmfs, computes their joint distribution, and returns a joint pmf of each possible pair of values and its probability.

```
def MakeJoint(pmf1, pmf2):
    joint = Joint()
    for v1, p1 in pmf1.Items():
        for v2, p2 in pmf2.Items():
            joint.Set((v1, v2), p1 * p2)
    return joint
```

This function assumes that the two distributions are independent.

Figure 12-5 shows the joint posterior distribution of p_correct for Alice and Bob. The diagonal line indicates the part of the space where p_correct is the same for Alice and Bob. To the right of this line, Alice is more prepared; to the left, Bob is more prepared.

In TopLevel.Update, when we compute the likelihoods of *A* and *B*, we add up the probability mass on each side of this line. For the cells that fall on the line, we add up the total mass and split it between *A* and *B*.

The process we used in this chapter—estimating nuisance parameters in order to evaluate the likelihood of competing hypotheses—is a common Bayesian approach to problems like this.

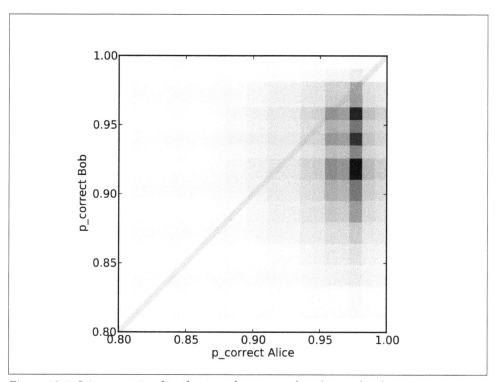

Figure 12-5. Joint posterior distribution of p_correct for Alice and Bob.

CHAPTER 13
Simulation

In this chapter I describe my solution to a problem posed by a patient with a kidney tumor. I think the problem is important and relevant to patients with these tumors and doctors treating them.

And I think the solution is interesting because, although it is a Bayesian approach to the problem, the use of Bayes's theorem is implicit. I present the solution and my code; at the end of the chapter I will explain the Bayesian part.

If you want more technical detail than I present here, you can read my paper on this work at *http://arxiv.org/abs/1203.6890*.

The Kidney Tumor problem

I am a frequent reader and occasional contributor to the online statistics forum at *http://reddit.com/r/statistics*. In November 2011, I read the following message:

> "I have Stage IV Kidney Cancer and am trying to determine if the cancer formed before I retired from the military. ... Given the dates of retirement and detection is it possible to determine when there was a 50/50 chance that I developed the disease? Is it possible to determine the probability on the retirement date? My tumor was 15.5 cm x 15 cm at detection. Grade II."

I contacted the author of the message and got more information; I learned that veterans get different benefits if it is "more likely than not" that a tumor formed while they were in military service (among other considerations).

Because renal tumors grow slowly, and often do not cause symptoms, they are sometimes left untreated. As a result, doctors can observe the rate of growth for untreated tumors by comparing scans from the same patient at different times. Several papers have reported these growth rates.

I collected data from a paper by Zhang et al[1]. I contacted the authors to see if I could get raw data, but they refused on grounds of medical privacy. Nevertheless, I was able to extract the data I needed by printing one of their graphs and measuring it with a ruler.

They report growth rates in reciprocal doubling time (RDT), which is in units of doublings per year. So a tumor with $RDT = 1$ doubles in volume each year; with $RDT = 2$ it quadruples in the same time, and with $RDT = -1$, it halves. Figure 13-1 shows the distribution of RDT for 53 patients.

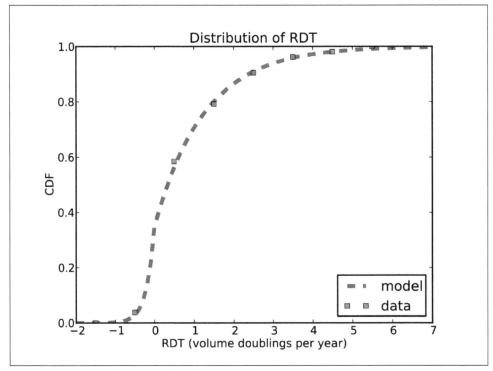

Figure 13-1. CDF of RDT in doublings per year.

The squares are the data points from the paper; the line is a model I fit to the data. The positive tail fits an exponential distribution well, so I used a mixture of two exponentials.

---

1 Zhang et al, Distribution of Renal Tumor Growth Rates Determined by Using Serial Volumetric CT Measurements, January 2009 *Radiology*, 250, 137-144.

A simple model

It is usually a good idea to start with a simple model before trying something more challenging. Sometimes the simple model is sufficient for the problem at hand, and if not, you can use it to validate the more complex model.

For my simple model, I assume that tumors grow with a constant doubling time, and that they are three-dimensional in the sense that if the maximum linear measurement doubles, the volume is multiplied by eight.

I learned from my correspondent that the time between his discharge from the military and his diagnosis was 3291 days (about 9 years). So my first calculation was, "If this tumor grew at the median rate, how big would it have been at the date of discharge?"

The median volume doubling time reported by Zhang et al is 811 days. Assuming 3-dimensional geometry, the doubling time for a linear measure is three times longer.

```
# time between discharge and diagnosis, in days
interval = 3291.0

# doubling time in linear measure is doubling time in volume * 3
dt = 811.0 * 3

# number of doublings since discharge
doublings = interval / dt

# how big was the tumor at time of discharge (diameter in cm)
d1 = 15.5
d0 = d1 / 2.0 ** doublings
```

You can download the code in this chapter from *http://thinkbayes.com/kidney.py*. For more information see "Working with the code" on page xi.

The result, d0, is about 6 cm. So if this tumor formed after the date of discharge, it must have grown substantially faster than the median rate. Therefore I concluded that it is "more likely than not" that this tumor formed before the date of discharge.

In addition, I computed the growth rate that would be implied if this tumor had formed after the date of discharge. If we assume an initial size of 0.1 cm, we can compute the number of doublings to get to a final size of 15.5 cm:

```
# assume an initial linear measure of 0.1 cm
d0 = 0.1
d1 = 15.5

# how many doublings would it take to get from d0 to d1
doublings = log2(d1 / d0)

# what linear doubling time does that imply?
```

```
    dt = interval / doublings

    # compute the volumetric doubling time and RDT
    vdt = dt / 3
    rdt = 365 / vdt
```

`dt` is linear doubling time, so `vdt` is volumetric doubling time, and `rdt` is reciprocal doubling time.

The number of doublings, in linear measure, is 7.3, which implies an RDT of 2.4. In the data from Zhang et al, only 20% of tumors grew this fast during a period of observation. So again, I concluded that is "more likely than not" that the tumor formed prior to the date of discharge.

These calculations are sufficient to answer the question as posed, and on behalf of my correspondent, I wrote a letter explaining my conclusions to the Veterans' Benefit Administration.

Later I told a friend, who is an oncologist, about my results. He was surprised by the growth rates observed by Zhang et al, and by what they imply about the ages of these tumors. He suggested that the results might be interesting to researchers and doctors.

But in order to make them useful, I wanted a more general model of the relationship between age and size.

A more general model

Given the size of a tumor at time of diagnosis, it would be most useful to know the probability that the tumor formed before any given date; in other words, the distribution of ages.

To find it, I run simulations of tumor growth to get the distribution of size conditioned on age. Then we can use a Bayesian approach to get the distribution of age conditioned on size.

The simulation starts with a small tumor and runs these steps:

1. Choose a growth rate from the distribution of RDT.
2. Compute the size of the tumor at the end of an interval.
3. Record the size of the tumor at each interval.
4. Repeat until the tumor exceeds the maximum relevant size.

For the initial size I chose 0.3 cm, because carcinomas smaller than that are less likely to be invasive and less likely to have the blood supply needed for rapid growth (see http://en.wikipedia.org/wiki/Carcinoma_in_situ).

I chose an interval of 245 days (about 8 months) because that is the median time between measurements in the data source.

For the maximum size I chose 20 cm. In the data source, the range of observed sizes is 1.0 to 12.0 cm, so we are extrapolating beyond the observed range at each end, but not by far, and not in a way likely to have a strong effect on the results.

The simulation is based on one big simplification: the growth rate is chosen independently during each interval, so it does not depend on age, size, or growth rate during previous intervals.

In "Serial Correlation" on page 152 I review these assumptions and consider more detailed models. But first let's look at some examples.

Figure 13-2 shows the size of simulated tumors as a function of age. The dashed line at 10 cm shows the range of ages for tumors at that size: the fastest-growing tumor gets there in 8 years; the slowest takes more than 35.

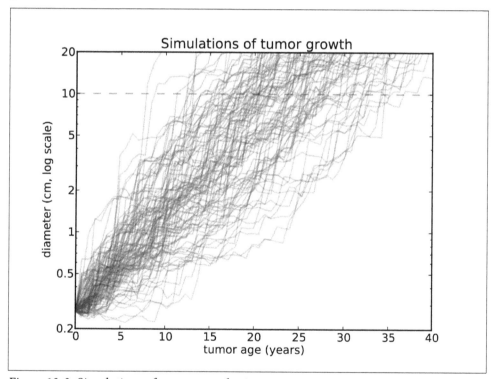

Figure 13-2. Simulations of tumor growth, size vs. time.

I am presenting results in terms of linear measurements, but the calculations are in terms of volume. To convert from one to the other, again, I use the volume of a sphere with the given diameter.

Implementation

Here is the kernel of the simulation:

```
def MakeSequence(rdt_seq, v0=0.01, interval=0.67, vmax=Volume(20.0)):
    seq = v0,
    age = 0

    for rdt in rdt_seq:
        age += interval
        final, seq = ExtendSequence(age, seq, rdt, interval)
        if final > vmax:
            break

    return seq
```

`rdt_seq` is an iterator that yields random values from the CDF of growth rate. `v0` is the initial volume in mL. `interval` is the time step in years. `vmax` is the final volume corresponding to a linear measurement of 20 cm.

`Volume` converts from linear measurement in cm to volume in mL, based on the simplification that the tumor is a sphere:

```
def Volume(diameter, factor=4*math.pi/3):
    return factor * (diameter/2.0)**3
```

`ExtendSequence` computes the volume of the tumor at the end of the interval.

```
def ExtendSequence(age, seq, rdt, interval):
    initial = seq[-1]
    doublings = rdt * interval
    final = initial * 2**doublings
    new_seq = seq + (final,)
    cache.Add(age, new_seq, rdt)

    return final, new_seq
```

`age` is the age of the tumor at the end of the interval. `seq` is a tuple that contains the volumes so far. `rdt` is the growth rate during the interval, in doublings per year. `interval` is the size of the time step in years.

The return values are `final`, the volume of the tumor at the end of the interval, and `new_seq`, a new tuple containing the volumes in `seq` plus the new volume `final`.

`Cache.Add` records the age and size of each tumor at the end of each interval, as explained in the next section.

Caching the joint distribution

Here's how the cache works.

```
class Cache(object):

    def __init__(self):
        self.joint = thinkbayes.Joint()
```

joint is a joint Pmf that records the frequency of each age-size pair, so it approximates the joint distribution of age and size.

At the end of each simulated interval, ExtendSequence calls Add:

```
# class Cache

    def Add(self, age, seq):
        final = seq[-1]
        cm = Diameter(final)
        bucket = round(CmToBucket(cm))
        self.joint.Incr((age, bucket))
```

Again, age is the age of the tumor, and seq is the sequence of volumes so far.

Before adding the new data to the joint distribution, we use Diameter to convert from volume to diameter in centimeters:

```
def Diameter(volume, factor=3/math.pi/4, exp=1/3.0):
    return 2 * (factor * volume) ** exp
```

And CmToBucket to convert from centimeters to a discrete bucket number:

```
def CmToBucket(x, factor=10):
    return factor * math.log(x)
```

The buckets are equally spaced on a log scale. Using factor=10 yields a reasonable number of buckets; for example, 1 cm maps to bucket 0 and 10 cm maps to bucket 23.

After running the simulations, we can plot the joint distribution as a pseudocolor plot, where each cell represents the number of tumors observed at a given size-age pair. Figure 13-3 shows the joint distribution after 1000 simulations.

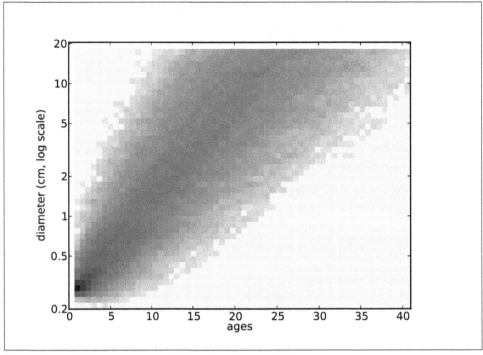

Figure 13-3. Joint distribution of age and tumor size.

Conditional distributions

By taking a vertical slice from the joint distribution, we can get the distribution of sizes for any given age. By taking a horizontal slice, we can get the distribution of ages conditioned on size.

Here's the code that reads the joint distribution and builds the conditional distribution for a given size.

```
# class Cache

    def ConditionalCdf(self, bucket):
        pmf = self.joint.Conditional(0, 1, bucket)
        cdf = pmf.MakeCdf()
        return cdf
```

bucket is the integer bucket number corresponding to tumor size. Joint.Conditional computes the PMF of age conditioned on bucket. The result is the CDF of age conditioned on bucket.

Figure 13-4 shows several of these CDFs, for a range of sizes. To summarize these distributions, we can compute percentiles as a function of size.

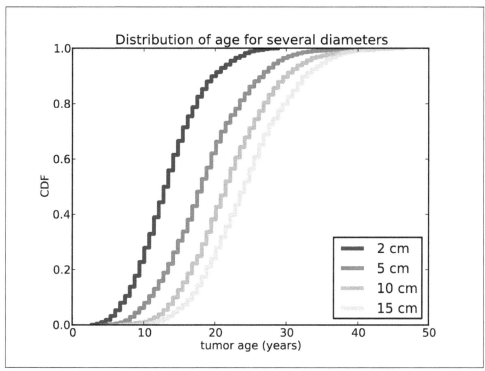

Figure 13-4. Distributions of age, conditioned on size.

```
percentiles = [95, 75, 50, 25, 5]

for bucket in cache.GetBuckets():
    cdf = ConditionalCdf(bucket)
    ps = [cdf.Percentile(p) for p in percentiles]
```

Figure 13-5 shows these percentiles for each size bucket. The data points are computed from the estimated joint distribution. In the model, size and time are discrete, which contributes numerical errors, so I also show a least squares fit for each sequence of percentiles.

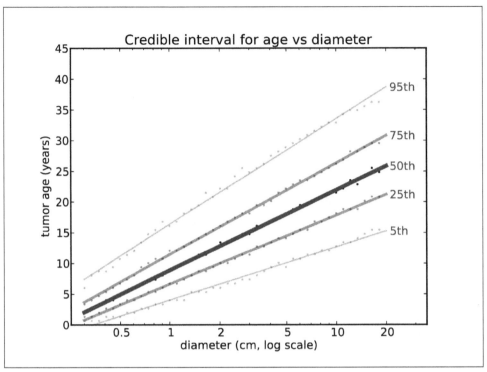

Figure 13-5. Percentiles of tumor age as a function of size.

Serial Correlation

The results so far are based on a number of modeling decisions; let's review them and consider which ones are the most likely sources of error:

- To convert from linear measure to volume, we assume that tumors are approximately spherical. This assumption is probably fine for tumors up to a few centimeters, but not for very large tumors.
- The distribution of growth rates in the simulations are based on a continuous model we chose to fit the data reported by Zhang et al, which is based on 53 patients. The fit is only approximate and, more importantly, a larger sample would yield a different distribution.
- The growth model does not take into account tumor subtype or grade; this assumption is consistent with the conclusion of Zhang et al: "Growth rates in renal tumors of different sizes, subtypes and grades represent a wide range and overlap substantially." But with a larger sample, a difference might become apparent.

- The distribution of growth rate does not depend on the size of the tumor. This assumption would not be realistic for very small and very large tumors, whose growth is limited by blood supply.

  But tumors observed by Zhang et al ranged from 1 to 12 cm, and they found no statistically significant relationship between size and growth rate. So if there is a relationship, it is likely to be weak, at least in this size range.

- In the simulations, growth rate during each interval is independent of previous growth rates. In reality it is plausible that tumors that have grown quickly in the past are more likely to grow quickly. In other words, there is probably a serial correlation in growth rate.

Of these, the first and last seem the most problematic. I'll investigate serial correlation first, then come back to spherical geometry.

To simulate correlated growth, I wrote a generator[2] that yields a correlated series from a given Cdf. Here's how the algorithm works:

1. Generate correlated values from a Gaussian distribution. This is easy to do because we can compute the distribution of the next value conditioned on the previous value.
2. Transform each value to its cumulative probability using the Gaussian CDF.
3. Transform each cumulative probability to the corresponding value using the given Cdf.

Here's what that looks like in code:

```
def CorrelatedGenerator(cdf, rho):
    x = random.gauss(0, 1)
    yield Transform(x)

    sigma = math.sqrt(1 - rho**2);
    while True:
        x = random.gauss(x * rho, sigma)
        yield Transform(x)
```

cdf is the desired Cdf; rho is the desired correlation. The values of x are Gaussian; Transform converts them to the desired distribution.

The first value of x is Gaussian with mean 0 and standard deviation 1. For subsequent values, the mean and standard deviation depend on the previous value. Given the previous x, the mean of the next value is x * rho, and the variance is 1 - rho**2.

Transform maps from each Gaussian value, x, to a value from the given Cdf, y.

---

[2] If you are not familiar with Python generators, see *http://wiki.python.org/moin/Generators*.

```
def Transform(x):
    p = thinkbayes.GaussianCdf(x)
    y = cdf.Value(p)
    return y
```

`GaussianCdf` computes the CDF of the standard Gaussian distribution at x, returning a cumulative probability. `Cdf.Value` maps from a cumulative probability to the corresponding value in `cdf`.

Depending on the shape of `cdf`, information can be lost in transformation, so the actual correlation might be lower than `rho`. For example, when I generate 10000 values from the distribution of growth rates with `rho=0.4`, the actual correlation is 0.37. But since we are guessing at the right correlation anyway, that's close enough.

Remember that `MakeSequence` takes an iterator as an argument. That interface allows it to work with different generators:

```
iterator = UncorrelatedGenerator(cdf)
seq1 = MakeSequence(iterator)

iterator = CorrelatedGenerator(cdf, rho)
seq2 = MakeSequence(iterator)
```

In this example, `seq1` and `seq2` are drawn from the same distribution, but the values in `seq1` are uncorrelated and the values in `seq2` are correlated with a coefficient of approximately `rho`.

Now we can see what effect serial correlation has on the results; the following table shows percentiles of age for a 6 cm tumor, using the uncorrelated generator and a correlated generator with target $\rho = 0.4$.

Table 13-1. Percentiles of tumor age conditioned on size.

Serial Correlation	Diameter (cm)	Percentiles of age				
		5th	25th	50th	75th	95th
0.0	6.0	10.7	15.4	19.5	23.5	30.2
0.4	6.0	9.4	15.4	20.8	26.2	36.9

Correlation makes the fastest growing tumors faster and the slowest slower, so the range of ages is wider. The difference is modest for low percentiles, but for the 95th percentile it is more than 6 years. To compute these percentiles precisely, we would need a better estimate of the actual serial correlation.

However, this model is sufficient to answer the question we started with: given a tumor with a linear dimension of 15.5 cm, what is the probability that it formed more than 8 years ago?

Here's the code:

```
# class Cache

    def ProbOlder(self, cm, age):
        bucket = CmToBucket(cm)
        cdf = self.ConditionalCdf(bucket)
        p = cdf.Prob(age)
        return 1-p
```

cm is the size of the tumor; age is the age threshold in years. ProbOlder converts size to a bucket number, gets the Cdf of age conditioned on bucket, and computes the probability that age exceeds the given value.

With no serial correlation, the probability that a 15.5 cm tumor is older than 8 years is 0.999, or almost certain. With correlation 0.4, faster-growing tumors are more likely, but the probability is still 0.995. Even with correlation 0.8, the probability is 0.978.

Another likely source of error is the assumption that tumors are approximately spherical. For a tumor with linear dimensions 15.5 x 15 cm, this assumption is probably not valid. If, as seems likely, a tumor this size is relatively flat, it might have the same volume as a 6 cm sphere. With this smaller volume and correlation 0.8, the probability of age greater than 8 is still 95%.

So even taking into account modeling errors, it is unlikely that such a large tumor could have formed less than 8 years prior to the date of diagnosis.

Discussion

Well, we got through a whole chapter without using Bayes's theorem or the Suite class that encapsulates Bayesian updates. What happened?

One way to think about Bayes's theorem is as an algorithm for inverting conditional probabilities. Given $p(B|A)$, we can compute $p(A|B)$, provided we know $p(A)$ and $p(B)$. Of course this algorithm is only useful if, for some reason, it is easier to compute $p(B|A)$ than $p(A|B)$.

In this example, it is. By running simulations, we can estimate the distribution of size conditioned on age, or p(*size*|*age*). But it is harder to get the distribution of age conditioned on size, or p(*age*|*size*). So this seems like a perfect opportunity to use Bayes's theorem.

The reason I didn't is computational efficiency. To estimate p(*size*|*age*) for any given size, you have to run a lot of simulations. Along the way, you end up computing p(*size*|*age*) for a lot of sizes. In fact, you end up computing the entire joint distribution of size and age, p(*size*, *age*).

And once you have the joint distribution, you don't really need Bayes's theorem, you can extract p(*age*|*size*) by taking slices from the joint distribution, as demonstrated in `ConditionalCdf`.

So we side-stepped Bayes, but he was with us in spirit.

CHAPTER 14
A Hierarchical Model

The Geiger counter problem

I got the idea for the following problem from Tom Campbell-Ricketts, author of the Maximum Entropy blog at *http://maximum-entropy-blog.blogspot.com*. And he got the idea from E.T. Jaynes, author of the classic *Probability Theory: The Logic of Science*:

> Suppose that a radioactive source emits particles toward a Geiger counter at an average rate of r particles per second, but the counter only registers a fraction, f, of the particles that hit it. If f is 10% and the counter registers 15 particles in a one second interval, what is the posterior distribution of n, the actual number of particles that hit the counter, and r, the average rate particles are emitted?

To get started on a problem like this, think about the chain of causation that starts with the parameters of the system and ends with the observed data:

1. The source emits particles at an average rate, r.
2. During any given second, the source emits n particles toward the counter.
3. Out of those n particles, some number, k, get counted.

The probability that an atom decays is the same at any point in time, so radioactive decay is well modeled by a Poisson process. Given r, the distribution of n is Poisson distribution with parameter r.

And if we assume that the probability of detection for each particle is independent of the others, the distribution of k is the binomial distribution with parameters n and f.

Given the parameters of the system, we can find the distribution of the data. So we can solve what is called the **forward problem**.

Now we want to go the other way: given the data, we want the distribution of the parameters. This is called the **inverse problem**. And if you can solve the forward problem, you can use Bayesian methods to solve the inverse problem.

Start simple

Let's start with a simple version of the problem where we know the value of r. We are given the value of f, so all we have to do is estimate n.

I define a Suite called `Detector` that models the behavior of the detector and estimates n.

```
class Detector(thinkbayes.Suite):

    def __init__(self, r, f, high=500, step=1):
        pmf = thinkbayes.MakePoissonPmf(r, high, step=step)
        thinkbayes.Suite.__init__(self, pmf, name=r)
        self.r = r
        self.f = f
```

If the average emission rate is r particles per second, the distribution of n is Poisson with parameter r. `high` and `step` determine the upper bound for n and the step size between hypothetical values.

Now we need a likelihood function:

```
# class Detector

    def Likelihood(self, data, hypo):
        k = data
        n = hypo
        p = self.f

        return thinkbayes.EvalBinomialPmf(k, n, p)
```

`data` is the number of particles detected, and `hypo` is the hypothetical number of particles emitted, n.

If there are actually n particles, and the probability of detecting any one of them is f, the probability of detecting k particles is given by the binomial distribution.

That's it for the Detector. We can try it out for a range of values of r:

```
f = 0.1
k = 15

for r in [100, 250, 400]:
    suite = Detector(r, f, step=1)
    suite.Update(k)
    print suite.MaximumLikelihood()
```

Figure 14-1 shows the posterior distribution of n for several given values of r.

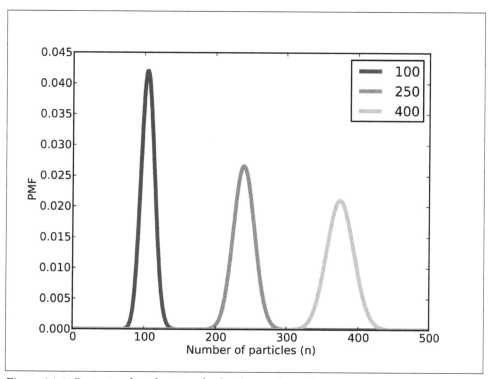

Figure 14-1. Posterior distribution of n for three values of r.

Make it hierarchical

In the previous section, we assume *r* is known. Now let's relax that assumption. I define another Suite, called `Emitter`, that models the behavior of the emitter and estimates *r*:

```
class Emitter(thinkbayes.Suite):

    def __init__(self, rs, f=0.1):
        detectors = [Detector(r, f) for r in rs]
        thinkbayes.Suite.__init__(self, detectors)
```

`rs` is a sequence of hypothetical value for *r*. `detectors` is a sequence of Detector objects, one for each value of *r*. The values in the Suite are Detectors, so Emitter is a **meta-Suite**; that is, a Suite that contains other Suites as values.

To update the Emitter, we have to compute the likelihood of the data under each hypothetical value of *r*. But each value of *r* is represented by a Detector that contains a range of values for *n*.

To compute the likelihood of the data for a given Detector, we loop through the values of n and add up the total probability of k. That's what `SuiteLikelihood` does:

```
# class Detector

    def SuiteLikelihood(self, data):
        total = 0
        for hypo, prob in self.Items():
            like = self.Likelihood(data, hypo)
            total += prob * like
        return total
```

Now we can write the Likelihood function for the Emitter:

```
# class Emitter

    def Likelihood(self, data, hypo):
        detector = hypo
        like = detector.SuiteLikelihood(data)
        return like
```

Each `hypo` is a Detector, so we can invoke `SuiteLikelihood` to get the likelihood of the data under the hypothesis.

After we update the Emitter, we have to update each of the Detectors, too.

```
# class Emitter

    def Update(self, data):
        thinkbayes.Suite.Update(self, data)

        for detector in self.Values():
            detector.Update()
```

A model like this, with multiple levels of Suites, is called **hierarchical**.

A little optimization

You might recognize `SuiteLikelihood`; we saw it in "Making a fair comparison" on page 124. At the time, I pointed out that we didn't really need it, because the total probability computed by `SuiteLikelihood` is exactly the normalizing constant computed and returned by `Update`.

So instead of updating the Emitter and then updating the Detectors, we can do both steps at the same time, using the result from `Detector.Update` as the likelihood of Emitter.

Here's the streamlined version of `Emitter.Likelihood`:

```
# class Emitter

    def Likelihood(self, data, hypo):
        return hypo.Update(data)
```

And with this version of `Likelihood` we can use the default version of `Update`. So this version has fewer lines of code, and it runs faster because it does not compute the normalizing constant twice.

Extracting the posteriors

After we update the Emitter, we can get the posterior distribution of r by looping through the Detectors and their probabilities:

```
# class Emitter

    def DistOfR(self):
        items = [(detector.r, prob) for detector, prob in self.Items()]
        return thinkbayes.MakePmfFromItems(items)
```

`items` is a list of values of r and their probabilities. The result is the Pmf of r.

To get the posterior distribution of n, we have to compute the mixture of the Detectors. We can use `thinkbayes.MakeMixture`, which takes a meta-Pmf that maps from each distribution to its probability. And that's exactly what the Emitter is:

```
# class Emitter

    def DistOfN(self):
        return thinkbayes.MakeMixture(self)
```

Figure 14-2 shows the results. Not surprisingly, the most likely value for n is 150. Given f and n, the expected count is $k = fn$, so given f and k, the expected value of n is k/f, which is 150.

And if 150 particles are emitted in one second, the most likely value of r is 150 particles per second. So the posterior distribution of r is also centered on 150.

The posterior distributions of r and n are similar; the only difference is that we are slightly less certain about n. In general, we can be more certain about the long-range emission rate, r, than about the number of particles emitted in any particular second, n.

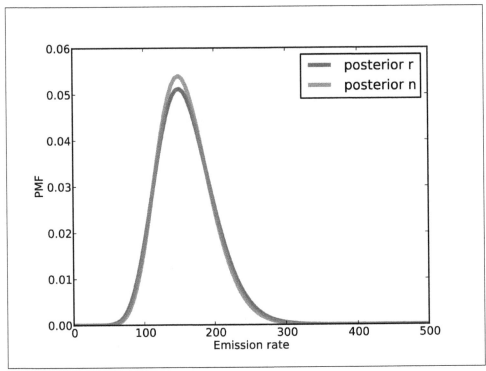

Figure 14-2. Posterior distributions of n and r.

You can download the code in this chapter from *http://thinkbayes.com/jaynes.py*. For more information see "Working with the code" on page xi.

Discussion

The Geiger counter problem demonstrates the connection between causation and hierarchical modeling. In the example, the emission rate r has a causal effect on the number of particles, n, which has a causal effect on the particle count, k.

The hierarchical model reflects the structure of the system, with causes at the top and effects at the bottom.

1. At the top level, we start with a range of hypothetical values for r.
2. For each value of r, we have a range of values for n, and the prior distribution of n depends on r.
3. When we update the model, we go bottom-up. We compute a posterior distribution of n for each value of r, then compute the posterior distribution of r.

So causal information flows down the hierarchy, and inference flows up.

Exercises

Exercise 14-1.

This exercise is also inspired by an example in Jaynes, *Probability Theory*.

Suppose you buy a mosquito trap that is supposed to reduce the population of mosquitoes near your house. Each week, you empty the trap and count the number of mosquitoes captured. After the first week, you count 30 mosquitoes. After the second week, you count 20 mosquitoes. Estimate the percentage change in the number of mosquitoes in your yard.

To answer this question, you have to make some modeling decisions. Here are some suggestions:

- Suppose that each week a large number of mosquitoes, N, is bred in a wetland near your home.
- During the week, some fraction of them, f_1, wander into your yard, and of those some fraction, f_2, are caught in the trap.
- Your solution should take into account your prior belief about how much N is likely to change from one week to the next. You can do that by adding a level to the hierarchy to model the percent change in N.

CHAPTER 15
Dealing with Dimensions

Belly button bacteria

Belly Button Biodiversity 2.0 (BBB2) is a nation-wide citizen science project with the goal of identifying bacterial species that can be found in human navels (*http://bbdata.yourwildlife.org*). The project might seem whimsical, but it is part of an increasing interest in the human microbiome, the set of microorganisms that live on human skin and parts of the body.

In their pilot study, BBB2 researchers collected swabs from the navels of 60 volunteers, used multiplex pyrosequencing to extract and sequence fragments of 16S rDNA, then identified the species or genus the fragments came from. Each identified fragment is called a "read."

We can use these data to answer several related questions:

- Based on the number of species observed, can we estimate the total number of species in the environment?
- Can we estimate the prevalence of each species; that is, the fraction of the total population belonging to each species?
- If we are planning to collect additional samples, can we predict how many new species we are likely to discover?
- How many additional reads are needed to increase the fraction of observed species to a given threshold?

These questions make up what is called the **Unseen Species problem**.

Lions and tigers and bears

I'll start with a simplified version of the problem where we know that there are exactly three species. Let's call them lions, tigers and bears. Suppose we visit a wild animal preserve and see 3 lions, 2 tigers and one bear.

If we have an equal chance of observing any animal in the preserve, the number of each species we see is governed by the multinomial distribution. If the prevalence of lions and tigers and bears is p_lion and p_tiger and p_bear, the likelihood of seeing 3 lions, 2 tigers and one bear is proportional to

```
p_lion**3 * p_tiger**2 * p_bear**1
```

An approach that is tempting, but not correct, is to use beta distributions, as in "The beta distribution" on page 37, to describe the prevalence of each species separately. For example, we saw 3 lions and 3 non-lions; if we think of that as 3 "heads" and 3 "tails," then the posterior distribution of p_lion is:

```
beta = thinkbayes.Beta()
beta.Update((3, 3))
print beta.MaximumLikelihood()
```

The maximum likelihood estimate for p_lion is the observed rate, 50%. Similarly the MLEs for p_tiger and p_bear are 33% and 17%.

But there are two problems:

1. We have implicitly used a prior for each species that is uniform from 0 to 1, but since we know that there are three species, that prior is not correct. The right prior should have a mean of 1/3, and there should be zero likelihood that any species has a prevalence of 100%.

2. The distributions for each species are not independent, because the prevalences have to add up to 1. To capture this dependence, we need a joint distribution for the three prevalences.

We can use a Dirichlet distribution to solve both of these problems (see *http://en.wikipedia.org/wiki/Dirichlet_distribution*). In the same way we used the beta distribution to describe the distribution of bias for a coin, we can use a Dirichlet distribution to describe the joint distribution of p_lion, p_tiger and p_bear.

The Dirichlet distribution is the multi-dimensional generalization of the beta distribution. Instead of two possible outcomes, like heads and tails, the Dirichlet distribution handles any number of outcomes: in this example, three species.

If there are n outcomes, the Dirichlet distribution is described by n parameters, written α_1 through α_n.

Here's the definition, from `thinkbayes.py`, of a class that represents a Dirichlet distribution:

```
class Dirichlet(object):

    def __init__(self, n):
        self.n = n
        self.params = numpy.ones(n, dtype=numpy.int)
```

n is the number of dimensions; initially the parameters are all 1. I use a `numpy` array to store the parameters so I can take advantage of array operations.

Given a Dirichlet distribution, the marginal distribution for each prevalence is a beta distribution, which we can compute like this:

```
    def MarginalBeta(self, i):
        alpha0 = self.params.sum()
        alpha = self.params[i]
        return Beta(alpha, alpha0-alpha)
```

`i` is the index of the marginal distribution we want. `alpha0` is the sum of the parameters; `alpha` is the parameter for the given species.

In the example, the prior marginal distribution for each species is `Beta(1, 2)`. We can compute the prior means like this:

```
    dirichlet = thinkbayes.Dirichlet(3)
    for i in range(3):
        beta = dirichlet.MarginalBeta(i)
        print beta.Mean()
```

As expected, the prior mean prevalence for each species is 1/3.

To update the Dirichlet distribution, we add the observations to the parameters like this:

```
    def Update(self, data):
        m = len(data)
        self.params[:m] += data
```

Here `data` is a sequence of counts in the same order as `params`, so in this example, it should be the number of lions, tigers and bears.

`data` can be shorter than `params`; in that case there are some species that have not been observed.

Here's code that updates `dirichlet` with the observed data and computes the posterior marginal distributions.

```
data = [3, 2, 1]
dirichlet.Update(data)

for i in range(3):
    beta = dirichlet.MarginalBeta(i)
    pmf = beta.MakePmf()
    print i, pmf.Mean()
```

Figure 15-1 shows the results. The posterior mean prevalences are 44%, 33%, and 22%.

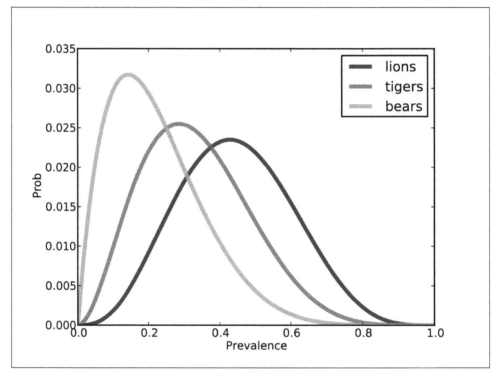

Figure 15-1. Distribution of prevalences for three species.

The hierarchical version

We have solved a simplified version of the problem: if we know how many species there are, we can estimate the prevalence of each.

Now let's get back to the original problem, estimating the total number of species. To solve this problem I'll define a meta-Suite, which is a Suite that contains other Suites as hypotheses. In this case, the top-level Suite contains hypotheses about the number of species; the bottom level contains hypotheses about prevalences.

Here's the class definition:

```
class Species(thinkbayes.Suite):

    def __init__(self, ns):
        hypos = [thinkbayes.Dirichlet(n) for n in ns]
        thinkbayes.Suite.__init__(self, hypos)
```

__init__ takes a list of possible values for n and makes a list of Dirichlet objects.

Here's the code that creates the top-level suite:

```
ns = range(3, 30)
suite = Species(ns)
```

ns is the list of possible values for n. We have seen 3 species, so there have to be at least that many. I chose an upper bound that seems reasonable, but we will check later that the probability of exceeding this bound is low. And at least initially we assume that any value in this range is equally likely.

To update a hierarchical model, you have to update all levels. Usually you have to update the bottom level first and work up, but in this case we can update the top level first:

```
#class Species

    def Update(self, data):
        thinkbayes.Suite.Update(self, data)
        for hypo in self.Values():
            hypo.Update(data)
```

Species.Update invokes Update in the parent class, then loops through the sub-hypotheses and updates them.

Now all we need is a likelihood function:

```
# class Species

    def Likelihood(self, data, hypo):
        dirichlet = hypo
        like = 0
        for i in range(1000):
            like += dirichlet.Likelihood(data)

        return like
```

data is a sequence of observed counts; hypo is a Dirichlet object. Species.Likelihood calls Dirichlet.Likelihood 1000 times and returns the total.

Why call it 1000 times? Because Dirichlet.Likelihood doesn't actually compute the likelihood of the data under the whole Dirichlet distribution. Instead, it draws one sample from the hypothetical distribution and computes the likelihood of the data under the sampled set of prevalences.

Here's what it looks like:

```
# class Dirichlet

    def Likelihood(self, data):
        m = len(data)
        if self.n < m:
            return 0

        x = data
        p = self.Random()
        q = p[:m]**x
        return q.prod()
```

The length of `data` is the number of species observed. If we see more species than we thought existed, the likelihood is 0.

Otherwise we select a random set of prevalences, p, and compute the multinomial PMF, which is

$$c_x p_1^{x_1} \cdots p_n^{x_n}$$

p_i is the prevalence of the *i*th species, and x_i is the observed number. The first term, c_x, is the multinomial coefficient; I leave it out of the computation because it is a multiplicative factor that depends only on the data, not the hypothesis, so it gets normalized away (see *http://en.wikipedia.org/wiki/Multinomial_distribution*).

m is the number of observed species. We only need the first m elements of p; for the others, x_i is 0, so $p_i^{x_i}$ is 1, and we can leave them out of the product.

Random sampling

There are two ways to generate a random sample from a Dirichlet distribution. One is to use the marginal beta distributions, but in that case you have to select one at a time and scale the rest so they add up to 1 (see *http://en.wikipedia.org/wiki/Dirichlet_distribution#Random_number_generation*).

A less obvious, but faster, way is to select values from n gamma distributions, then normalize by dividing through by the total. Here's the code:

```
# class Dirichlet

    def Random(self):
        p = numpy.random.gamma(self.params)
        return p / p.sum()
```

Now we're ready to look at some results. Here is the code that extracts the posterior distribution of n:

```
def DistOfN(self):
    pmf = thinkbayes.Pmf()
    for hypo, prob in self.Items():
        pmf.Set(hypo.n, prob)
    return pmf
```

`DistOfN` iterates through the top-level hypotheses and accumulates the probability of each n.

Figure 15-2 shows the result. The most likely value is 4. Values from 3 to 7 are reasonably likely; after that the probabilities drop off quickly. The probability that there are 29 species is low enough to be negligible; if we chose a higher bound, we would get nearly the same result.

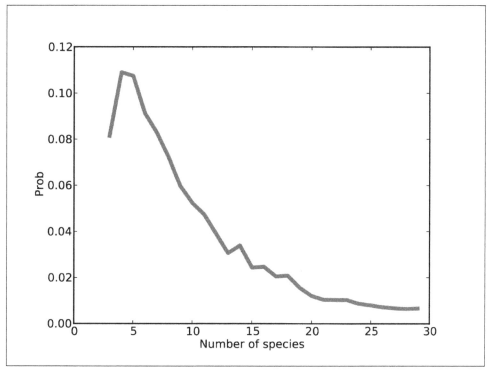

Figure 15-2. Posterior distribution of n.

Remember that this result is based on a uniform prior for n. If we have background information about the number of species in the environment, we might choose a different prior.

Optimization

I have to admit that I am proud of this example. The Unseen Species problem is not easy, and I think this solution is simple and clear, and takes surprisingly few lines of code (about 50 so far).

The only problem is that it is slow. It's good enough for the example with only 3 observed species, but not good enough for the belly button data, with more than 100 species in some samples.

The next few sections present a series of optimizations we need to make this solution scale. Before we get into the details, here's a road map.

- The first step is to recognize that if we update the Dirichlet distributions with the same data, the first m parameters are the same for all of them. The only difference is the number of hypothetical unseen species. So we don't really need n Dirichlet objects; we can store the parameters in the top level of the hierarchy. Species2 implements this optimization.

- Species2 also uses the same set of random values for all of the hypotheses. This saves time generating random values, but it has a second benefit that turns out to be more important: by giving all hypotheses the same selection from the sample space, we make the comparison between the hypotheses more fair, so it takes fewer iterations to converge.

- Even with these changes there is a major performance problem. As the number of observed species increases, the array of random prevalences gets bigger, and the chance of choosing one that is approximately right becomes small. So the vast majority of iterations yield small likelihoods that don't contribute much to the total, and don't discriminate between hypotheses.

  The solution is to do the updates one species at a time. Species4 is a simple implementation of this strategy using Dirichlet objects to represent the sub-hypotheses.

- Finally, Species5 combines the sub-hypotheses into the top level and uses numpy array operations to speed things up.

If you are not interested in the details, feel free to skip to "The belly button data" on page 178 where we look at results from the belly button data.

Collapsing the hierarchy

All of the bottom-level Dirichlet distributions are updated with the same data, so the first m parameters are the same for all of them. We can eliminate them and merge the parameters into the top-level suite. Species2 implements this optimization:

```
class Species2(object):

    def __init__(self, ns):
        self.ns = ns
        self.probs = numpy.ones(len(ns), dtype=numpy.double)
        self.params = numpy.ones(self.high, dtype=numpy.int)
```

`ns` is the list of hypothetical values for n; `probs` is the list of corresponding probabilities. And `params` is the sequence of Dirichlet parameters, initially all 1.

`Species2.Update` updates both levels of the hierarchy: first the probability for each value of n, then the Dirichlet parameters:

```
# class Species2

    def Update(self, data):
        like = numpy.zeros(len(self.ns), dtype=numpy.double)
        for i in range(1000):
            like += self.SampleLikelihood(data)

        self.probs *= like
        self.probs /= self.probs.sum()

        m = len(data)
        self.params[:m] += data
```

`SampleLikelihood` returns an array of likelihoods, one for each value of n. `like` accumulates the total likelihood for 1000 samples. `self.probs` is multiplied by the total likelihood, then normalized. The last two lines, which update the parameters, are the same as in `Dirichlet.Update`.

Now let's look at `SampleLikelihood`. There are two opportunities for optimization here:

- When the hypothetical number of species, n, exceeds the observed number, m, we only need the first m terms of the multinomial PMF; the rest are 1.
- If the number of species is large, the likelihood of the data might be too small for floating-point (see "Underflow" on page 111). So it is safer to compute log-likelihoods.

Again, the multinomial PMF is

$$c_x p_1^{x_1} \ldots p_n^{x_n}$$

So the log-likelihood is

$$\log c_x + x_1 \log p_1 + \cdots + x_n \log p_n$$

which is fast and easy to compute. Again, c_x it is the same for all hypotheses, so we can drop it. Here's the code:

```
# class Species2

    def SampleLikelihood(self, data):
        gammas = numpy.random.gamma(self.params)

        m = len(data)
        row = gammas[:m]
        col = numpy.cumsum(gammas)

        log_likes = []
        for n in self.ns:
            ps = row / col[n-1]
            terms = data * numpy.log(ps)
            log_like = terms.sum()
            log_likes.append(log_like)

        log_likes -= numpy.max(log_likes)
        likes = numpy.exp(log_likes)

        coefs = [thinkbayes.BinomialCoef(n, m) for n in self.ns]
        likes *= coefs

        return likes
```

gammas is an array of values from a gamma distribution; its length is the largest hypothetical value of n. row is just the first m elements of gammas; since these are the only elements that depend on the data, they are the only ones we need.

For each value of n we need to divide row by the total of the first n values from gamma. cumsum computes these cumulative sums and stores them in col.

The loop iterates through the values of n and accumulates a list of log-likelihoods.

Inside the loop, ps contains the row of probabilities, normalized with the appropriate cumulative sum. terms contains the terms of the summation, $x_i \log p_i$, and log_like contains their sum.

After the loop, we want to convert the log-likelihoods to linear likelihoods, but first it's a good idea to shift them so the largest log-likelihood is 0; that way the linear likelihoods are not too small (see "Underflow" on page 111).

Finally, before we return the likelihood, we have to apply a correction factor, which is the number of ways we could have observed these m species, if the total number of species is n. BinomialCoefficient computes "n choose m", which is written $\binom{n}{m}$.

As often happens, the optimized version is less readable and more error-prone than the original. But that's one reason I think it is a good idea to start with the simple version; we can use it for regression testing. I plotted results from both versions and confirmed that they are approximately equal, and that they converge as the number of iterations increases.

One more problem

There's more we could do to optimize this code, but there's another problem we need to fix first. As the number of observed species increases, this version gets noisier and takes more iterations to converge on a good answer.

The problem is that if the prevalences we choose from the Dirichlet distribution, the ps, are not at least approximately right, the likelihood of the observed data is close to zero and almost equally bad for all values of n. So most iterations don't provide any useful contribution to the total likelihood. And as the number of observed species, m, gets large, the probability of choosing ps with non-negligible likelihood gets small. Really small.

Fortunately, there is a solution. Remember that if you observe a set of data, you can update the prior distribution with the entire dataset, or you can break it up into a series of updates with subsets of the data, and the result is the same either way.

For this example, the key is to perform the updates one species at a time. That way when we generate a random set of ps, only one of them affects the computed likelihood, so the chance of choosing a good one is much better.

Here's a new version that updates one species at a time:

```
class Species4(Species):

    def Update(self, data):
        m = len(data)

        for i in range(m):
            one = numpy.zeros(i+1)
            one[i] = data[i]
            Species.Update(self, one)
```

This version inherits __init__ from Species, so it represents the hypotheses as a list of Dirichlet objects (unlike Species2).

Update loops through the observed species and makes an array, one, with all zeros and one species count. Then it calls Update in the parent class, which computes the likelihoods and updates the sub-hypotheses.

So in the running example, we do three updates. The first is something like "I have seen three lions." The second is "I have seen two tigers and no additional lions." And the third is "I have seen one bear and no more lions and tigers."

Here's the new version of `Likelihood`:

```
# class Species4

    def Likelihood(self, data, hypo):
        dirichlet = hypo
        like = 0
        for i in range(self.iterations):
            like += dirichlet.Likelihood(data)

            # correct for the number of unseen species the new one
            # could have been
            m = len(data)
            num_unseen = dirichlet.n - m + 1
            like *= num_unseen

        return like
```

This is almost the same as `Species.Likelihood`. The difference is the factor, `num_unseen`. This correction is necessary because each time we see a species for the first time, we have to consider that there were some number of other unseen species that we might have seen. For larger values of n there are more unseen species that we could have seen, which increases the likelihood of the data.

This is a subtle point and I have to admit that I did not get it right the first time. But again I was able to validate this version by comparing it to the previous versions.

We're not done yet

Performing the updates one species at a time solves one problem, but it creates another. Each update takes time proportional to *km*, where *k* is the number of hypotheses and *m* is the number of observed species. So if we do *m* updates, the total run time is proportional to km^2.

But we can speed things up using the same trick we used in "Collapsing the hierarchy" on page 172: we'll get rid of the Dirichlet objects and collapse the two levels of the hierarchy into a single object. So here's yet another version of `Species`:

```
class Species5(Species2):

    def Update(self, data):
        m = len(data)
        for i in range(m):
            self.UpdateOne(i+1, data[i])
            self.params[i] += data[i]
```

This version inherits `__init__` from `Species2`, so it uses `ns` and `probs` to represent the distribution of n, and `params` to represent the parameters of the Dirichlet distribution.

Update is similar to what we saw in the previous section. It loops through the observed species and calls UpdateOne:

```
# class Species5

    def UpdateOne(self, i, count):
        likes = numpy.zeros(len(self.ns), dtype=numpy.double)
        for i in range(self.iterations):
            likes += self.SampleLikelihood(i, count)

        unseen_species = [n-i+1 for n in self.ns]
        likes *= unseen_species

        self.probs *= likes
        self.probs /= self.probs.sum()
```

This function is similar to Species2.Update, with two changes:

- The interface is different. Instead of the whole dataset, we get i, the index of the observed species, and count, how many of that species we've seen.
- We have to apply a correction factor for the number of unseen species, as in Species4.Likelihood. The difference here is that we update all of the likelihoods at once with array multiplication.

Finally, here's SampleLikelihood:

```
# class Species5

    def SampleLikelihood(self, i, count):
        gammas = numpy.random.gamma(self.params)

        sums = numpy.cumsum(gammas)[self.ns[0]-1:]

        ps = gammas[i-1] / sums
        log_likes = numpy.log(ps) * count

        log_likes -= numpy.max(log_likes)
        likes = numpy.exp(log_likes)

        return likes
```

This is similar to Species2.SampleLikelihood; the difference is that each update only includes a single species, so we don't need a loop.

The runtime of this function is proportional to the number of hypotheses, k. It runs m times, so the run time of the update is proportional to km. And the number of iterations we need to get an accurate result is usually small.

The belly button data

That's enough about lions and tigers and bears. Let's get back to belly buttons. To get a sense of what the data look like, consider subject B1242, whose sample of 400 reads yielded 61 species with the following counts:

```
92, 53, 47, 38, 15, 14, 12, 10, 8, 7, 7, 5, 5,
4, 4, 4, 4, 4, 4, 4, 3, 3, 3, 3, 3, 3, 3, 2, 2, 2, 2,
1, 1, 1, 1, 1, 1, 1, 1, 1, 1, 1, 1, 1, 1, 1, 1, 1, 1,
1, 1, 1, 1, 1, 1, 1, 1, 1, 1, 1, 1, 1
```

There are a few dominant species that make up a large fraction of the whole, but many species that yielded only a single read. The number of these "singletons" suggests that there are likely to be at least a few unseen species.

In the example with lions and tigers, we assume that each animal in the preserve is equally likely to be observed. Similarly, for the belly button data, we assume that each bacterium is equally likely to yield a read.

In reality, each step in the data-collection process might introduce biases. Some species might be more likely to be picked up by a swab, or to yield identifiable amplicons. So when we talk about the prevalence of each species, we should remember this source of error.

I should also acknowledge that I am using the term "species" loosely. First, bacterial species are not well defined. Second, some reads identify a particular species, others only identify a genus. To be more precise, I should say "operational taxonomic unit", or OTU.

Now let's process some of the belly button data. I define a class called `Subject` to represent information about each subject in the study:

```
class Subject(object):

    def __init__(self, code):
        self.code = code
        self.species = []
```

Each subject has a string code, like "B1242", and a list of (count, species name) pairs, sorted in increasing order by count. `Subject` provides several methods to make it easy to access these counts and species names. You can see the details in *http://thinkbayes.com/species.py*. For more information see "Working with the code" on page xi.

`Subject` provides a method named `Process` that creates and updates a `Species5` suite, which represents the distributions of n and the prevalences.

And `Suite2` provides `DistOfN`, which returns the posterior distribution of n.

```
# class Suite2

    def DistN(self):
        items = zip(self.ns, self.probs)
        pmf = thinkbayes.MakePmfFromItems(items)
        return pmf
```

Figure 15-3 shows the distribution of n for subject B1242. The probability that there are exactly 61 species, and no unseen species, is nearly zero. The most likely value is 72, with 90% credible interval 66 to 79. At the high end, it is unlikely that there are as many as 87 species.

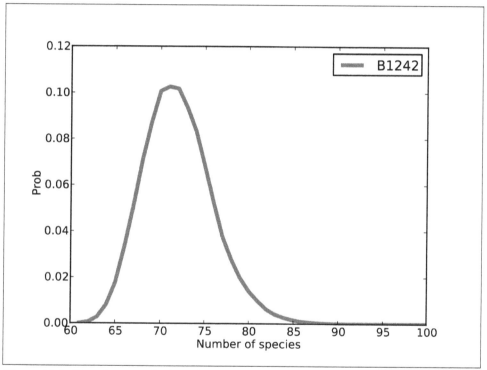

Figure 15-3. Distribution of n for subject B1242.

Next we compute the posterior distribution of prevalence for each species. `Species2` provides `DistOfPrevalence`:

```
# class Species2

    def DistOfPrevalence(self, index):
        metapmf = thinkbayes.Pmf()

        for n, prob in zip(self.ns, self.probs):
            beta = self.MarginalBeta(n, index)
            pmf = beta.MakePmf()
            metapmf.Set(pmf, prob)

        mix = thinkbayes.MakeMixture(metapmf)
        return metapmf, mix
```

`index` indicates which species we want. For each `n`, we have a different posterior distribution of prevalence.

The loop iterates through the possible values of `n` and their probabilities. For each value of `n` it gets a Beta object representing the marginal distribution for the indicated species. Remember that Beta objects contain the parameters `alpha` and `beta`; they don't have values and probabilities like a Pmf, but they provide `MakePmf`, which generates a discrete approximation to the continuous beta distribution.

`metapmf` is a meta-Pmf that contains the distributions of prevalence, conditioned on `n`. `MakeMixture` combines the meta-Pmf into `mix`, which combines the conditional distributions into a single distribution of prevalence.

Figure 15-4 shows results for the five species with the most reads. The most prevalent species accounts for 23% of the 400 reads, but since there are almost certainly unseen species, the most likely estimate for its prevalence is 20%, with 90% credible interval between 17% and 23%.

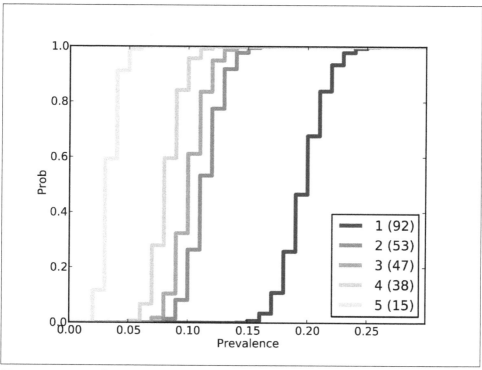

Figure 15-4. Distribution of prevalences for subject B1242.

Predictive distributions

I introduced the hidden species problem in the form of four related questions. We have answered the first two by computing the posterior distribution for n and the prevalence of each species.

The other two questions are:

- If we are planning to collect additional reads, can we predict how many new species we are likely to discover?
- How many additional reads are needed to increase the fraction of observed species to a given threshold?

To answer predictive questions like this we can use the posterior distributions to simulate possible future events and compute predictive distributions for the number of species, and fraction of the total, we are likely to see.

The kernel of these simulations looks like this:

1. Choose n from its posterior distribution.
2. Choose a prevalence for each species, including possible unseen species, using the Dirichlet distribution.
3. Generate a random sequence of future observations.
4. Compute the number of new species, num_new, as a function of the number of additional reads, k.
5. Repeat the previous steps and accumulate the joint distribution of num_new and k.

And here's the code. RunSimulation runs a single simulation:

```
# class Subject

    def RunSimulation(self, num_reads):
        m, seen = self.GetSeenSpecies()
        n, observations = self.GenerateObservations(num_reads)

        curve = []
        for k, obs in enumerate(observations):
            seen.add(obs)

            num_new = len(seen) - m
            curve.append((k+1, num_new))

        return curve
```

num_reads is the number of additional reads to simulate. m is the number of seen species, and seen is a set of strings with a unique name for each species. n is a random value from the posterior distribution, and observations is a random sequence of species names.

Each time through the loop, we add the new observation to seen and record the number of reads and the number of new species so far.

The result of RunSimulation is a **rarefaction curve**, represented as a list of pairs with the number of reads and the number of new species.

Before we see the results, let's look at GetSeenSpecies and GenerateObservations.

```
#class Subject

    def GetSeenSpecies(self):
        names = self.GetNames()
        m = len(names)
        seen = set(SpeciesGenerator(names, m))
        return m, seen
```

GetNames returns the list of species names that appear in the data files, but for many subjects these names are not unique. So I use SpeciesGenerator to extend each name with a serial number:

```
def SpeciesGenerator(names, num):
    i = 0
    for name in names:
        yield '%s-%d' % (name, i)
        i += 1

    while i < num:
        yield 'unseen-%d' % i
        i += 1
```

Given a name like Corynebacterium, SpeciesGenerator yields Corynebacterium-1. When the list of names is exhausted, it yields names like unseen-62.

Here is GenerateObservations:

```
# class Subject

    def GenerateObservations(self, num_reads):
        n, prevalences = self.suite.SamplePosterior()

        names = self.GetNames()
        name_iter = SpeciesGenerator(names, n)

        d = dict(zip(name_iter, prevalences))
        cdf = thinkbayes.MakeCdfFromDict(d)
        observations = cdf.Sample(num_reads)

        return n, observations
```

Again, num_reads is the number of additional reads to generate. n and prevalences are samples from the posterior distribution.

cdf is a Cdf object that maps species names, including the unseen, to cumulative probabilities. Using a Cdf makes it efficient to generate a random sequence of species names.

Finally, here is Species2.SamplePosterior:

```
    def SamplePosterior(self):
        pmf = self.DistOfN()
        n = pmf.Random()
        prevalences = self.SamplePrevalences(n)
        return n, prevalences
```

And SamplePrevalences, which generates a sample of prevalences conditioned on n:

```
# class Species2

    def SamplePrevalences(self, n):
        params = self.params[:n]
        gammas = numpy.random.gamma(params)
        gammas /= gammas.sum()
        return gammas
```

We saw this algorithm for generating random values from a Dirichlet distribution in "Random sampling" on page 170.

Figure 15-5 shows 100 simulated rarefaction curves for subject B1242. The curves are "jittered;" that is, I shifted each curve by a random offset so they would not all overlap. By inspection we can estimate that after 400 more reads we are likely to find 2–6 new species.

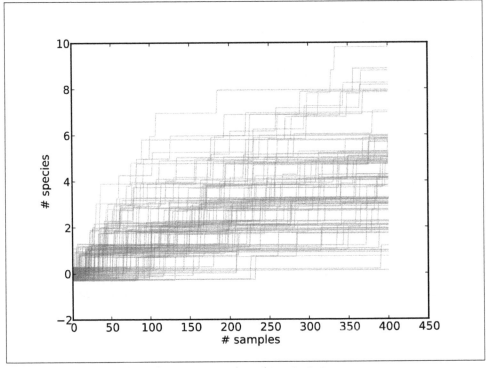

Figure 15-5. Simulated rarefaction curves for subject B1242.

Joint posterior

We can use these simulations to estimate the joint distribution of num_new and k, and from that we can get the distribution of num_new conditioned on any value of k.

```
def MakeJointPredictive(curves):
    joint = thinkbayes.Joint()
    for curve in curves:
        for k, num_new in curve:
            joint.Incr((k, num_new))
    joint.Normalize()
    return joint
```

MakeJointPredictive makes a Joint object, which is a Pmf whose values are tuples.

curves is a list of rarefaction curves created by RunSimulation. Each curve contains a list of pairs of k and num_new.

The resulting joint distribution is a map from each pair to its probability of occurring. Given the joint distribution, we can use Joint.Conditional get the distribution of num_new conditioned on k (see "Conditional distributions" on page 101).

Subject.MakeConditionals takes a list of ks and computes the conditional distribution of num_new for each k. The result is a list of Cdf objects.

```
def MakeConditionals(curves, ks):
    joint = MakeJointPredictive(curves)

    cdfs = []
    for k in ks:
        pmf = joint.Conditional(1, 0, k)
        pmf.name = 'k=%d' % k
        cdf = pmf.MakeCdf()
        cdfs.append(cdf)

    return cdfs
```

Figure 15-6 shows the results. After 100 reads, the median predicted number of new species is 2; the 90% credible interval is 0 to 5. After 800 reads, we expect to see 3 to 12 new species.

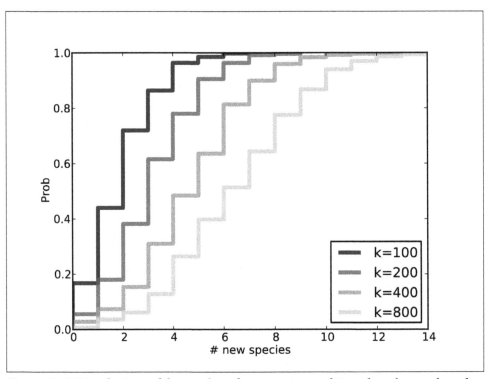

Figure 15-6. Distributions of the number of new species conditioned on the number of additional reads.

Coverage

The last question we want to answer is, "How many additional reads are needed to increase the fraction of observed species to a given threshold?"

To answer this question, we need a version of `RunSimulation` that computes the fraction of observed species rather than the number of new species.

```
# class Subject

    def RunSimulation(self, num_reads):
        m, seen = self.GetSeenSpecies()
        n, observations = self.GenerateObservations(num_reads)

        curve = []
        for k, obs in enumerate(observations):
            seen.add(obs)

            frac_seen = len(seen) / float(n)
            curve.append((k+1, frac_seen))

        return curve
```

Next we loop through each curve and make a dictionary, d, that maps from the number of additional reads, k, to a list of `fracs`; that is, a list of values for the coverage achieved after k reads.

```
def MakeFracCdfs(self, curves):
    d = {}
    for curve in curves:
        for k, frac in curve:
            d.setdefault(k, []).append(frac)

    cdfs = {}
    for k, fracs in d.iteritems():
        cdf = thinkbayes.MakeCdfFromList(fracs)
        cdfs[k] = cdf

    return cdfs
```

Then for each value of k we make a Cdf of `fracs`; this Cdf represents the distribution of coverage after k reads.

Remember that the CDF tells you the probability of falling below a given threshold, so the *complementary* CDF tells you the probability of exceeding it. Figure 15-7 shows complementary CDFs for a range of values of k.

To read this figure, select the level of coverage you want to achieve along the *x*-axis. As an example, choose 90%.

Now you can read up the chart to find the probability of achieving 90% coverage after k reads. For example, with 200 reads, you have about a 40% chance of getting 90% coverage. With 1000 reads, you have a 90% chance of getting 90% coverage.

With that, we have answered the four questions that make up the unseen species problem. To validate the algorithms in this chapter with real data, I had to deal with a few more details. But this chapter is already too long, so I won't discuss them here.

You can read about the problems, and how I addressed them, at *http://allendowney.blogspot.com/2013/05/belly-button-biodiversity-end-game.html*.

You can download the code in this chapter from *http://thinkbayes.com/species.py*. For more information see "Working with the code" on page xi.

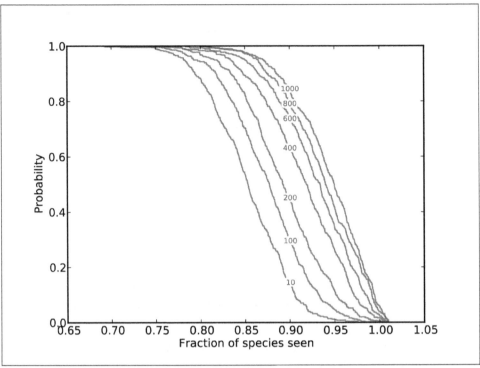

Figure 15-7. Complementary CDF of coverage for a range of additional reads.

Discussion

The Unseen Species problem is an area of active research, and I believe the algorithm in this chapter is a novel contribution. So in fewer than 200 pages we have made it from the basics of probability to the research frontier. I'm very happy about that.

My goal for this book is to present three related ideas:

- **Bayesian thinking:** The foundation of Bayesian analysis is the idea of using probability distributions to represent uncertain beliefs, using data to update those distributions, and using the results to make predictions and inform decisions.
- **A computational approach:** The premise of this book is that it is easier to understand Bayesian analysis using computation rather than math, and easier to implement Bayesian methods with reusable building blocks that can be rearranged to solve real-world problems quickly.
- **Iterative modeling:** Most real-world problems involve modeling decisions and trade-offs between realism and complexity. It is often impossible to know ahead of time what factors should be included in the model and which can be abstrac-

ted away. The best approach is to iterate, starting with simple models and adding complexity gradually, using each model to validate the others.

These ideas are versatile and powerful; they are applicable to problems in every area of science and engineering, from simple examples to topics of current research.

If you made it this far, you should be prepared to apply these tools to new problems relevant to your work. I hope you find them useful; let me know how it goes!

Index

A
ABC, 115
abstract type, 18, 56
Anaconda, xi
Approximate Bayesian Computation, 115
arrival rate, 87
Axtell, Robert, 24

B
bacteria, 165
Bayes factor, 43, 123-124, 134
Bayesian framework, 13
Bayes's theorem, 3
    derivation, 3
    odds form, 42
Behavioral Risk Factor Surveillance System, 107
belly button, 165
Bernoulli process, 69
beta distribution, 37, 166
Beta object, 37, 180
biased coin, 123
binomial coefficient, 174
binomial distribution, 132, 157, 158
binomial likelihood function, 37
biodiversity, 165
bogus, 110, 124
Boston, 79
Boston Bruins, 67
BRFSS, 107, 115
bucket, 149
bus stop problem, 76

C
cache, 115, 148

calibration, 136
Campbell-Ricketts, Tom, 157
carcinoma, 146
causation, 157, 162
CDC, 107
Cdf, 26, 52, 58, 84, 183
Centers for Disease Control, 107
central credible interval, 103
classical estimation, 109
clone, xi
coefficient of variation, 108
coin toss, 1
collectively exhaustive, 6
College Board, 130
complementary CDF, 187
concrete type, 18, 56
conditional distribution, 101, 105, 146, 150, 156, 185
conditional probability, 1
conjoint probability, 2
conjugate prior, 37
conjunction, 3
continuous distribution, 37
contributors, xiv
convergence, 35, 38
cookie problem, 3, 12, 42
cookie.py, 13
correlated random value, 153
coverage, 187
crank science, 107
credible interval, 26, 100
Cromwell, Oliver, 39
Cromwell's rule, 38
cumulative distribution function, 26, 84

cumulative probability, 153, 183
cumulative sum, 174

D

Davidson-Pilon, Cameron, 54
decision analysis, 53, 62, 66, 91
degree of belief, 1
density, 55, 57, 61, 109
dependence, 2, 101, 102
diachronic interpretation, 5
dice, 11, 19
Dice problem, 19
dice problem, 21
Dirichlet distribution, 166, 182
distribution, 11, 52, 65
    operations, 44
divide-and-conquer, 10
doubling time, 144
Dungeons and Dragons, 19, 44

E

efficacy, 134
enumeration, 44, 47
error, 58
ESP, 127
Euro problem, 31, 38, 115, 123
evidence, 4, 33, 43, 44, 100, 107, 123-124, 129
exception, 112
exponential distribution, 69, 73, 144
exponentiation, 47
extra-sensory perception, 127

F

fair coin, 123
fork, xi
forward problem, 157

G

gamma distribution, 170, 174
Gaussian distribution, 55, 56, 56, 59, 68, 108, 115, 117, 130, 136, 136, 153
Gaussian PDF, 56
Gee, Steve, 54
Geiger counter problem, 157, 162
generator, 153, 154, 183
German tank problem, 20, 28
Git, xi
GitHub, xi

growth rate, 152

H

heart attack, 1
height, 108
Heuer, Andreas, 69
hierarchical model, 160, 162, 168
Hoag, Dirk, 75
hockey, 67
horse racing, 42
Horsford, Eben Norton, 107
Hume, David, 128
hypothesis testing, 123

I

implementation, 18, 56
independence, 2, 7, 47, 48, 101, 102, 140, 147, 166
informative prior, 28
insect sampling problem, 76
installation, xii
inter-quartile range, 116
interface, 18, 56
intuition, 8
inverse problem, 158
IQR, 116
item response theory, 135
iterative modeling, 75
iterator, 148

J

Jaynes, E.T., 157
Joint, 100, 101, 103, 104, 108
joint distribution, 100, 105, 108, 140, 149, 150, 155, 166, 182, 185
Joint object, 185
Joint pmf, 96

K

KDE, 55, 57
kernel density estimation, 55, 57
Kidney tumor problem, 143

L

least squares fit, 151
light bulb problem, 76
likelihood, 5, 58, 85, 97, 99, 109, 121, 125, 158
Likelihood, 14

likelihood function, 21
likelihood ratio, 43, 124, 126, 134
linspace, 110
lions and tigers and bears, 166
locomotive problem, 20, 28, 115
log scale, 149
log transform, 111
log-likelihood, 113, 173, 174
logarithm, 111

M

M and M problem, 6, 16
MacKay, David, 31, 43, 94, 123
MakeMixture, 71, 73, 82, 90, 136, 180
marginal distribution, 100, 105, 167
matplotlib, xi
maximum, 47
maximum likelihood, 26, 33, 66, 103, 110, 113, 166
mean squared error, 23
Meckel, Johann, 107
median, 33
memoization, 114
meta-Pmf, 71, 73, 82, 90, 136, 180
meta-Suite, 159, 168
microbiome, 165
mixture, 50, 71, 73, 82, 90, 144, 180
modeling, ix, 28, 38, 75, 121, 129, 145, 146
modeling error, 134, 152, 155
Monty Hall problem, 7, 15
Mosteller, Frederick, 20
Mult, 12
multinomial coefficient, 170
multinomial distribution, 166, 170, 173
mutually exclusive, 6

N

National Hockey League, 67
navel, 165
NHL, 67
non-linear, 91
normal distribution, 56
normalize, 61
normalizing constant, 5, 7, 42, 160
nuisance parameter, 140
NumPy, xi
numpy, 57, 59, 63, 68, 88, 110, 136, 167, 170, 172-177

O

objectivity, 28
observer bias, 81, 91
odds, 41
Olin College, 79
Oliver's blood problem, 43
operational taxonomic unit, 178
optimization, 35, 113, 114, 161, 172
OTU, 178
overtime, 73

P

Paintball problem, 95
parameter, 37
PDF, 38, 68
Pdf, 55, 55
PEP 8, xii
percentile, 26, 150, 154
Pmf, 52, 55
Pmf class, 11
Pmf methods, 12
Poisson distribution, 69, 70, 71, 85, 158
Poisson process, x, 67, 68, 73, 76, 80, 157
posterior, 5
posterior distribution, 13, 33
power law, 24
predictive distribution, 76, 84, 86, 89, 139, 181
prevalence, 165, 168, 178
Price is Right, 53
prior, 5
prior distribution, 12, 23
Prob, 12
probability, 55
    conditional, 1
    conjoint, 2
probability density, 55
probability density function, 38, 55, 68
probability mass function, 11
process, 68
pseudocolor plot, 149
pyrosequencing, 165

R

radioactive decay, 157
random sample, 170, 183
rarefaction curve, 182, 185
raw score, 132
rDNA, 165

Red Line problem, 79
Reddit, 39, 143
regression testing, xi, 175, 176
renormalize, 13
repository, xi
robust estimation, 116

S

sample bias, 178
sample statistics, 116
SAT, 129
scaled score, 130
SciPy, xi
scipy, 56, 57, 113
serial correlation, 153, 154
Showcase, 53
simulation, 44, 47, 50, 146, 148, 182
Sivia, D.S., 95
species, 165, 178
sphere, 147, 152
standardized test, 129
stick, 8
strafing speed, 97
subjective prior, 5
subjectivity, 28
sudden death, 73
suite, 6
Suite class, 16
summary statistic, 66, 116, 121
swamping the priors, 35, 38

switch, 8

T

table method, 6
template method pattern, 18
total probability, 6
triangle distribution, 34, 126
trigonometry, 97
tumor type, 152
tuple, 36

U

uncertainty, 89
underflow, 111, 173
uniform distribution, 31, 50, 82, 171
uninformative prior, 28
Unseen Species problem, 165
Update, 13

V

Vancouver Canucks, 67
Variability Hypothesis, 107
Veterans' Benefit Administration, 146
volume, 147

W

Weibull distribution, 77
word frequency, 11

About the Author

Allen Downey is a Professor of Computer Science at the Olin College of Engineering. He has taught computer science at Wellesley College, Colby College, and U.C. Berkeley. He has a PhD in Computer Science from U.C. Berkeley and Master's and Bachelor's degrees from MIT.

Colophon

The animal on the cover of *Think Bayes* is a red striped mullet (*Mullus surmuletus*). This species of goatfish can be found in the Mediterranean Sea, east North Atlantic Ocean, and the Black Sea. Known for its distinct striped first dorsal fin, the red striped mullet is a favored delicacy in the Mediterranean—along with its brother goatfish, *Mullus barbatus*, which has a first dorsal fin that is not striped. However, the red striped mullet tends to be more prized and is said to taste similar to oysters. Stories of ancient Romans rearing the red striped mullet in ponds, attending to, caressing, and even teaching them to feed at the sound of a bell. These fish, generally weighing in under two pounds even when farm-raised, were sometimes sold for their weight in silver.

When left to the wild, red mullets are small bottom-feeding fish with a distinct double beard—known as barbels—on its lower lip, which it uses to probe the ocean floor for food. Because the red striped mullet feed on sandy and rocky bottoms at shallower depths, its barbels are less sensitive than its deep water feeding brother, the *Mullus barbatus*.

The cover image is from *Meyers Kleines Lexicon*. The cover fonts are URW Typewriter and Guardian Sans. The text font is Adobe Minion Pro; the heading font is Adobe Myriad Condensed; and the code font is Dalton Maag's Ubuntu Mono.

Have it your way.

O'Reilly eBooks

- Lifetime access to the book when you buy through oreilly.com
- Provided in up to four, DRM-free file formats, for use on the devices of your choice: PDF, .epub, Kindle-compatible .mobi, and Android .apk
- Fully searchable, with copy-and-paste, and print functionality
- We also alert you when we've updated the files with corrections and additions.

oreilly.com/ebooks/

Safari Books Online

- Access the contents and quickly search over 7000 books on technology, business, and certification guides
- Learn from expert video tutorials, and explore thousands of hours of video on technology and design topics
- Download whole books or chapters in PDF format, at no extra cost, to print or read on the go
- Early access to books as they're being written
- Interact directly with authors of upcoming books
- Save up to 35% on O'Reilly print books

See the complete Safari Library at safaribooksonline.com

©2014 O'Reilly Media, Inc. O'Reilly logo is a registered trademark of O'Reilly Media, Inc. 14373

Get even more for your money.

Join the O'Reilly Community, and register the O'Reilly books you own. It's free, and you'll get:

- $4.99 ebook upgrade offer
- 40% upgrade offer on O'Reilly print books
- Membership discounts on books and events
- Free lifetime updates to ebooks and videos
- Multiple ebook formats, DRM FREE
- Participation in the O'Reilly community
- Newsletters
- Account management
- 100% Satisfaction Guarantee

Signing up is easy:

1. Go to: oreilly.com/go/register
2. Create an O'Reilly login.
3. Provide your address.
4. Register your books.

Note: English-language books only

To order books online:
oreilly.com/store

For questions about products or an order:
orders@oreilly.com

To sign up to get topic-specific email announcements and/or news about upcoming books, conferences, special offers, and new technologies:
elists@oreilly.com

For technical questions about book content:
booktech@oreilly.com

To submit new book proposals to our editors:
proposals@oreilly.com

O'Reilly books are available in multiple DRM-free ebook formats. For more information:
oreilly.com/ebooks

Lightning Source UK Ltd.
Milton Keynes UK
UKHW02f0327140718
325646UK00003BA/5/P

9 781449 370787